Advanced Materials and Techniques for Reinforced Concrete Structures

Advanced Materials and Techniques for Reinforced Concrete Structures

Mohamed A. El-Reedy, Ph.D.
Consultant Engineer
Cairo, Egypt

CRC Press
Taylor & Francis Group
Boca Raton   London   New York

CRC Press is an imprint of the
Taylor & Francis Group, an **informa** business

CRC Press
Taylor & Francis Group
6000 Broken Sound Parkway NW, Suite 300
Boca Raton, FL 33487-2742

Library of Congress Cataloging-in-Publication Data

El-Reedy, Mohamed A. (Mohamed Abdallah)
    Advanced materials and techniques for reinforced concrete structures / author,
Mohamed El-Reedy.
        p. cm.
    "A CRC title."
    Includes bibliographical references and index.
    ISBN 978-1-4200-8891-5 (hardcover : alk. paper)
    1. Reinforced concrete construction. 2. Reinforced concrete--Quality control. 3.
Buildings, Reinforced concrete. I. Title.

TA683.E495 2009
624.1'8341--dc22   2009016991

Visit the Taylor & Francis Web site at
http://www.taylorandfrancis.com

and the CRC Press Web site at
http://www.crcpress.com

This book is dedicated to the spirits of my mother and my father, my wife, and my children, Maey, Hisham, and Mayar.

Contents

Preface

Nowadays, there is a race among investors for constructing high-rise buildings due to high profits from the real estate business worldwide. Therefore, the major challenge for structural engineers is how to achieve the goal for the investors and the architects as well.

The aim of this book is to present the most advanced materials and construction techniques used in the reinforced concrete structure industry. Because advanced materials are growing parallel to the development of new structure systems, the advantages and disadvantages of the different structure systems are very important from durability, reliability, and construction points of view.

This book is intended to serve as a guide to junior and senior engineers who work in design, construction, and maintenance of reinforced concrete structures to assist them in choosing the most reasonable structure system, materials, method of construction, and maintenance plan.

Construction quality control is the main factor that allows a firm to be competitive in the market, so this book is written to be easy for structural engineers to understand the statistical parameters that govern quality control in concrete construction projects. Its aim is to avoid complicated statistical terms, and instead be more applicable by using strong theoretical background information to help analyze and meet concrete construction quality control criteria.

This book consists of three main approaches: the first chapters present ways to control a project to achieve the owner's target and discusses different loads that affect the buildings from the point of view of the international codes. In addition, the different structure systems that are used in traditional and high-rise buildings will be presented with their advantages and disadvantages to assist the designer in choosing the optimum structure system to meet stability, reliability, and architectural requirements.

The second main approach is to describe the traditional and most recent materials used in concrete technology such as high strength concrete, high performance concrete, and self-compacted concrete. Moreover, the book presents the modern techniques used in all construction stages in temperate and hot climate regions. After construction, the structural engineers may face many problems with the durability of the concrete, which can be achieved by using modern materials to protect steel bars from corrosion and avoid problems in countries with hot climates.

In the last chapter, the new approach for the integrity management system is presented by describing the advanced maintenance plan philosophy as risk-based for reinforced concrete structures. Recently, there is a movement toward maintaining the reliability of structures from safety and economic points of view by developing a structure integrity management system, which is also discussed in the final chapters.

This book provides a practical guide to advanced materials, design, and construction techniques in concrete structures for normal and high-rise buildings with different environmental conditions and the modern approach to a concrete structure maintenance plan.

Mohamed Abdallah El-Reedy
Cairo, Egypt
elreedyma@yahoo.com

The Author

Mohamed A. El-Reedy, PhD, pursued a career in structural engineering. His main area of research is the reliability of concrete and steel structures. He has provided consulting services to different engineering companies and oil and gas industries in Egypt and to international companies including the International Egyptian Oil Company (IEOC) and British Petroleum (BP). Moreover, he provides concrete and steel structure design packages for residential buildings, warehouses, telecommunication towers, and electrical projects of WorleyParsons Egypt. He has participated in liquefied natural gas and natural gas liquid projects with international engineering firms. Currently, Dr. El-Reedy is responsible for reliability, inspection, and maintenance strategies for onshore concrete structures and offshore steel structure platforms. He has managed these tasks for one hundred of these structures in the Gulf of Suez in the Red Sea.

Dr. El-Reedy has consulted with and trained executives at many organizations, including the Arabian American Oil Company (ARAMCO), British Petroleum (BP), Apachi, Abu Dhabi Marine Operating Company (ADMA), the Abu Dhabi National Oil Company and King Saudi's Interior Ministry, Qatar Telecom, the Egyptian General Petroleum Corporation, Saudi Arabia Basic Industries Corporation (SABIC), the Kuwait Petroleum Corporation, and Qatar Petrochemical Company (QAPCO). He has taught technical courses about repair and maintenance for reinforced concrete structures and advanced materials worldwide.

Dr. El-Reedy has written numerous publications and presented many papers at local and international conferences sponsored by the American Society of Civil Engineers, the American Society of Mechanical Engineers, the American Concrete Institute, the American Society for Testing and Materials, and the American Petroleum Institute. He has published many research papers in international technical journals and has authored four books about total quality management, quality management and quality assurance, economic management for engineering projects, and repair and protection of reinforced concrete structures. He earned a bachelor's degree from Cairo University in 1990, a master's degree in 1995, and a PhD from Cairo University in 2000.

1 Introduction

The civilization of any country is measured by its advanced techniques and advanced materials used in constructing buildings. Concrete is the main element for construction materials, and its development follows developments in engineering research.

Ancient Egyptians used concrete in buildings and temples along with crushed stone as an aggregate and clay as an adhesive. However, the Greeks used concrete in their buildings and called it *Santorin Tofa* (El-Arian and Atta 1974), and history mentions that the Romans used a concrete-like material called pozzolan.

After that, concrete disappeared for long time and then appeared again in the 18th century. Following are famous scientists who worked with concrete:

- John Semitone used it to construct the Ediston lighthouse.
- Joseph Parker researched stones and their uses in concrete.
- Odgar researched cement made from limestone and clay.
- Vicat researched using cement from limestone and clay.
- Joseph Espedin used Portland cement.

At the end of 19th century and the beginning of the 20th century, major changes affected the shapes of buildings as architectural engineers and builders changed their points of view to ideas previously used from the European Renaissance, such as using columns and arches. To consider the function required and to achieve the architectural intent economically, concrete was the best solution.

Nowadays reinforced concrete is the most important material in the construction industry and is used for different types of civil engineering projects such as tunnels, bridges, airports, drainage and hydraulic projects, and others. Research focuses on increasing concrete's strength and its performance to to match the varieties of applications.

Reinforced concrete is considered cheap when compared to other building materials, so it has been used for high-rise buildings for mega-projects and also for small projects such as one-story buildings, and all these projects are performed by contractors and engineers with different capabilities. From here one can conclude that concrete is used for projects of different scale with competent and noncompetent laborers and contractors, so there are many precautions in different codes to overcome these variables.

Due to expanding investment in real estate and industry and competition in constructing high-rise buildings worldwide, there is a trend to use modern materials and construction techniques that can match designers' visions. Moreover, modern research aims for new structure systems to accommodate the needs of high-rise buildings and enhance the architectural point of view.

Culture, social life, and economics vary widely among different countries, especially between the rich countries and poor countries or between developed countries and countries on their way to development. So, it is difficult to apply the specifications and codes for one country to another as the loads required depend on the lives of the people and the laws of each country. For example, in some countries it is easy to convert residential buildings to commercial buildings, but other countries prohibit this conversion, so the probability of failure is different from one country to another. Moreover, the dead load value depends on the construction quality. On the other hand, the competency of the engineers, supervisors, and laborers differs from one country to another and this affects the quality of the concrete, and thus the probability of producing a good quality concrete is different.

Therefore, the modern codes and the philosophy of the codes and specifications will be discussed to clarify the answers to the following questions:

- What is the probability of structure failure?
- What are the factors that affect this probability?

In addition, a comparison of strength, loads, and design factors is presented for different international codes.

The main challenge that faces the engineer working in the concrete industry is to increase the strength and performance of the concrete to enhance its durability along its lifetime. Therefore, high strength concrete (HSC), high performance concrete (HPC), and ultra high performance concrete (UHPC) are widely discussed in the literature. Moreover, studies in Japan focus on concrete without compaction, which is called self-compacted concrete (SCC).

All these types of modern concrete depend on using the new, advanced materials and knowing their performance very well. These materials are the mineral compounds such as fly ash, silica fume, and blast slag, and synthetic materials, such as the superplasticizer and different types of admixtures that enhance the concrete's properties.

On the other hand, there are many development techniques focused on providing materials and new methodologies to protect steel reinforcement bars from corrosion.

The concrete industry is spread worldwide from very cold countries to very hot countries, hence there are many practical ways to be used in providing new materials or new construction techniques to overcome the problems due to cold weather and hot weather. It is essential to clearly understand the problems in concrete construction in cold and hot weather, as it is a very important step to employ the new materials and new methods and techniques in construction procedures to overcome severe climatic conditions.

There are traditional structure systems and new systems in concrete structures, and each system has advantages and disadvantages. The selection of any system depends on knowing the differences among systems and to choose the best from an economic view and to match the characteristics with the architectural and end user's needs.

The history of building tall buildings stretches from 1880 until now, as stated by Bryan Smith and Alex Coull in 1991. Nowadays, the high-rise building is frequently in demand in city centers due to the increased value of land and as a prestige symbol for a commercial organization. Therefore, usually there are many studies to develop a structure system to reach the heights required. For high-rise buildings there are many structure systems, and combinations of systems are important to increase the height of the building while accommodating the lateral loads from wind and earthquake. All these system are discussed in Chapter 2.

Chapter 3 discusses the philosophy of the codes and the probability of failure. Moreover, in this chapter there is a comparison of concrete's strength from ACI, BS, and Egyptian codes of practice, and also a discussion of the differences among them in calculating the dead, live, wind, and earthquake loads.

Chapter 4 discusses concrete materials and different tests that control the quality of the concrete, and also presents new materials that can be used to enhance the concrete's strength and performance as well as the environment issues such as using recycled concrete.

Concrete design mix is different from one code to another, and every code has its basis for calculating the required concrete strength, so concrete needs to be clearly understood, as is presented in Chapter 5.

Chapter 6 discusses special concrete. It presents all the new, modern materials to be used to produce HPC, HSC, and SCC. Moreover, there are many development techniques to protect the steel reinforcement bars, and all the available practical methods on the market will be illustrated.

Besides the new materials, there are traditional methods of concrete construction and new methods to produce better quality. The road map to reach higher concrete quality in the construction process using the modern construction technique is presented in Chapter 7.

Recently, studies of the reliability of reinforced concrete structures have provided different techniques such as qualitative and quantitative risk assessment. These techniques are illustrated in the last chapter.

All the methods that will assist one to implement the maintenance plan from an economic point of view and the most economic method among steel protection alternatives will be discussed in the final chapter.

REFERENCES

El-Arian, A. A., and A. M. Atta. 1974. *Concrete technology.* World Book.
Smith, B. S., and A. Coull. 1991. *Tall building structures: analysis and design.* New York: John Wiley & Sons.

2 Reinforced Concrete Structures

2.1 INTRODUCTION

In general, engineering projects have more than one stage to pass through, and these stages are the idea, the feasibility study, feed engineering, detailed engineering and construction, and the last stage known as commissioning and startup.

The feasibility study stage mainly depends on an earlier economic study as in this phase the owner will decide whether a project is profitable or not and also if the profit is within limits.

The stage of feed engineering is the most critical for the project from an engineering point of view as in this phase the engineering office will define the structure system. The office experience in these types of projects and the main reason for project success is choosing the correct structure system in this phase.

2.2 FEED (PRELIMINARY) ENGINEERING

This stage is critical from an engineering view as the project's success depends on this stage. Due to this criticality we usually consult an international engineering office that has lots of experience in these types of projects, as these projects vary according to their function and many differences surround projects for the petroleum industry, hotels, and hospitals.

In the case of small projects such as housing, an administration building, or a small factory, the preliminary engineering will be to define the types and whether they will be reinforced concrete or steel structures. In the case of reinforced concrete structures, will they be traditional reinforced concrete, or precast or prestress concrete? In addition, preliminary engineering defines whether the project will use solid slab, flat slab, or hollow block.

The structure system may be beams and columns, frames, or shear wall. The study of all these alternatives depends on other factors such as the building location and the owner's requirements.

In the case of big projects such as tourist resorts, housing compounds, or petroleum projects this stage will be more complicated from the survey, planning, and study of the different roads needed to leveling the earth and performing the soil investigation. Foundation alternatives include shallow foundation or raft or deep foundation and whether it will require driven piles or rotating. Study the best in the market and use the most advanced techniques that match project requirements.

Based on the criticality of this phase to the project and the engineers' competency, in huge projects, the owner must have qualified engineers and a strong organization with the capability to follow up the preliminary engineering in pre-feed and feed engineering in an accurate manner to achieve the project target, and join the different disciplines such as civil, mechanical, electrical, and in some projects chemical studies as all these disciplines will interface at this stage.

The owner should prepare the statement of requirement (SOR) document that accurately covers all the requirements of the project such as the project's goals and the owner's requirements. The preparation of this document begins the circulation of important documents in the quality assurance system as this document must define all the owner's requirements, so the preparation of this document takes skill as it contains the project goals, target, owner recommendations and standards and specifications, and technical information such as the land location and coordinates, expected weather conditions, and the project lifetime.

As in the example of building a hotel, the number of rooms should be identified, as well as the size of the rooms and their level of finishing, and the required facilities such as a swimming pool, cinema, cafeteria, or game room.

In the case of a liquefied natural gas project, identify the gas quantity, specifications, and type, and the temperature and pressure and all the other technical data concerning the final product that will be exported. The very critical information in the document is the project lifeline as the study of the structure alternatives depends on this information.

When the engineering office receives the SOR document, they will reply with a basis of design (BOD) document. Using this document the engineering office defines the codes and specifications that will be used in the design along with the equations and theory in the design calculations, and the software that will be used in the structure analysis and design. Moreover, this document identifies the number of drawing copies that will be sent to the owner; the sizes of the drawings may be requested in the document, as well as third party requirements such as the geotechnical data and metocean data in offshore projects.

Now, we are in the stage of feed engineering so all the alternatives of the structure system will be on the table for discussion. In choosing the best structure type and suitable structure system, we must focus on the ways of maintaining it in the future. The location of the building and the environmental conditions surrounding the building will define the ways of protecting the structure during its lifetime and the types and frequency of maintenance needed. In some special structures we can use very expensive systems like the cathodic protection system to protect the steel bars. On the other hand, we can use different types of steel such as stainless bars, which are very expensive but will not need any maintenance in future.

The building's operation and maintenance affect the preliminary design, for example, in the case of a petrochemical plant project we must ask whether we can repair the water tank or perform maintenance or clean it? By answering this question, we can make a decision as to whether to construct spare tanks in case of emergency or maintenance, or additional tanks are not needed. All these alternatives and decisions must be made at this stage. This needs highly qualified engineers with strong experience in the field of the project, as correction of any mistake will cost more after commissioning and startup of the project.

The choice of the structure system depends on selecting the suitable economic system that matches with the project goals and the skills of the people who will perform the future maintenance.

2.3 DETAILED ENGINEERING

After completing this stage the owner will have a complete engineering package with construction drawings, project specifications, and material lists. This stage needs more time and resources to finish along with a good communication system and strong cooperation among different disciplines to achieve a successful project. The deliverables for this stage are the final construction drawings with complete specifications, including materials requirements and project specifications that will be included in the tender package.

2.4 DIFFERENT STRUCTURE SYSTEMS

Selection of the structure system depends on the architectural design and the loads that affect the building along its lifetime; moreover, the building's height and dimensions define the structure system that resists the wind and earthquake load in an economic and safe matter.

In this section, traditional and advanced structure systems are discussed by clearly presenting the advantages and disadvantages for every system. The choice among the different alternatives will be aided by knowing the various structure systems.

2.4.1 BEAM AND SLAB SYSTEM

This type of structure system is traditional, and this type of slab is used in most projects as it is easy to fabricate (Figures 2.1–2.3). All laborers have strong knowledge about this system because of its common use. From a design view construction is easy and is known by all engineers, so the probability of error is very small.

It is usually used in domestic buildings as the average room dimensions are 4 m and the live load is about 200 kg/m^2 so it is considered lighter weight than other systems. The maximum length of the small span is about 5–6 m and the slab thickness is often 100, 120, 140, or 150 mm and in some minor cases the slab thickness will reach 160 mm. If the calculations present the need for more thickness, it is better to select another system to be more economical.

2.4.2 HOLLOW BLOCK SLAB

This slab system depends on increasing the slab depth to increase its inertia against the bending momentum and decrease the weight of the slab itself (Figure 2.4).

This slab type has a minimum thickness of 250 mm, so it can accommodate high load or it can cover a higher span than what can be covered by the beam and slab system.

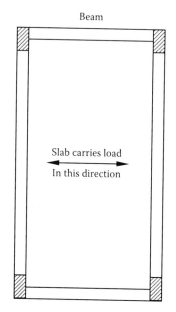

FIGURE 2.1   One-way slab.

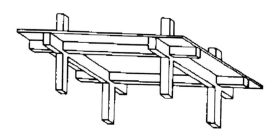

FIGURE 2.2   Two-way slab.

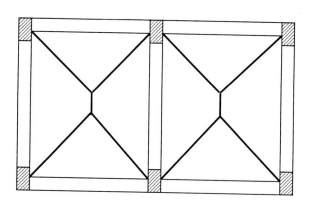

FIGURE 2.3   Load distribution in two-way slab.

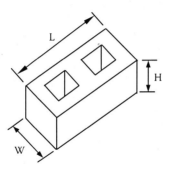

FIGURE 2.4   Cement block.

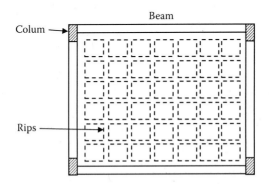

FIGURE 2.5   Hollow block slab.

The decrease of the dead load depends on placing hollow blocks as in Figure 2.5 or plastic blocks to reduce the slab self weight and the bigger thickness rips with a width of 100 mm to carry the stress load at the bottom of the slab, which is usually under tension stresses.

These types of slabs can cover dimensions of rooms 7 m by 7 m in residential buildings and can be used in administration buildings with heavier live loads and smaller dimensions.

When you choose this system, you must be sure that there is no vibration due to machines or rotating equipment.

In the hollow block system the rips can be in one direction or two directions according to the beam location and the columns.

Figure 2.5 shows the location of the rips in two directions and it can also be seen that execution of the hollow blocks of rips in one direction is easier than execution of the hollow block slabs with rips in two directions. Therefore, if you are concerned about the quality of the contractor or the competence of the supervising engineer on site, it is recommend that you avoid putting rips in two directions.

2.4.3   PAN JOIST FLOOR

This system is the same as the hollow block slab without filling between rips with cement bricks or plastic blocks.

FIGURE 2.6   Pan joist floor.

FIGURE 2.7   Pan joist floor system.

Figures 2.6 and 2.7 show the location of the rip, which is like a small beam with small width. One can see from the figures that this is reasonable for a special structure. The architectural solution is to cover the floor above using a false ceiling, so it is reasonable for commercial and administration buildings.

Figure 2.8 shows a section of a three-story building with a joist floor system and shows that the distance between joists is approximately 2.1 m.

2.4.4   FLAT SLAB

Figures 2.9 and 2.10 illustrate the difference between the slab and beam system and slab without beam and hollow block system. The slab without beam, which is

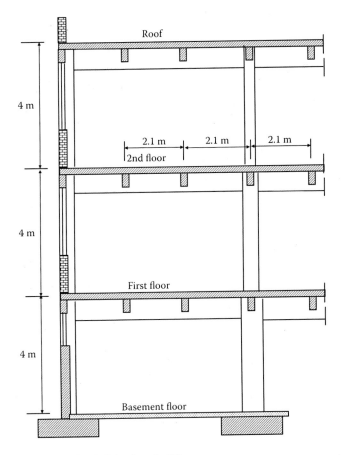

FIGURE 2.8 Cross section in joist floor building.

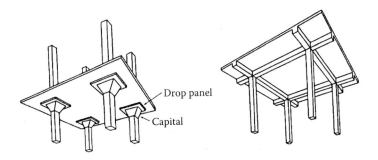

FIGURE 2.9 Slab with beam and flat slab with head.

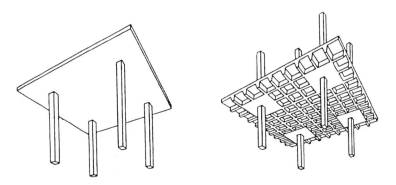

FIGURE 2.10   Flat slab without head and hollow block slab.

the flat slab, is used for span length up to 8 m and can carry high live load at the same time.

There are two methods for design: the first method needs certain dimensions between columns and certain distribution. The second method is to perform the structure analysis as in the frame design but it is less accurate. Now, using finite element methods with software that is available and easy to use, one can reach reasonable accuracy.

This system is used more widely now than previously, as the calculation is made easier by using the new software with higher accuracy. It is worth mentioning that the execution of the flat slab is easier than the beam and slab system and hollow block slab system. Unfortunately, the percentage of steel is higher and can reach to 250 kg/m³ compared with the slab beam system where the steel percentage is about 90–100 kg/m³.

The disadvantage of this system is that it is weaker in the case of earthquakes than the slab beam system, which provides rigidity against lateral forces; therefore, the designer in designing a flat slab system should consider earthquake load.

2.4.5   Prestressed Concrete

The use of prestressed concrete began in 1928 by Eugène Freyssinet. This method is the way to overcome the weakness of concrete to resist tensile stresses.

Concrete can resist high compressive stresses and very low tensile stresses. Due to this fact, cracks generally appear in the tension zone of reinforced concrete elements under working loads. For this reason, it is assumed in reinforced concrete design that concrete in tension does not act statically and steel reinforcement resists all the tensile stresses. The tension zone in reinforced concrete elements subjected to axial tension covers a whole section, and in elements subject to simple bending or eccentric forces it generally covers a big part of the cross section, which means that a large amount of the concrete used in such elements is cracked and not acting statically, although it adds more dead weight to the structure and consequently on the columns and foundation.

FIGURE 2.11   Pre-tensioned cables.

Tensioned steel cables are made from high-strength steel, and this tensioning to the steel bars results in compression stresses that resist the tensile stresses due to loads affecting the elements that provide flexural strength.

Prestressed concrete is used in beams, slabs, and bridges, especially in the case of large spans as it is more economical than regular reinforced concrete.

There are three ways to prestress concrete and they vary in the method of execution. These methods are

- Pre-tensioned concrete
- Bonded post-tensioned concrete
- Unbonded post-tensioned concrete

2.4.5.1   Pre-tensioned Concrete

Pre-tensioned concrete is cast around already tensioned tendons. This method produces a good bond between the tendon and concrete, which both protects the tendon from corrosion and allows for direct transfer of tension. The cured concrete adheres to the bars and when the tension is released it is transferred to the concrete as compression by static friction. However, this system requires strong anchoring points between which the tendon is to be stretched. The tendons are usually in a straight line. Thus, most pre-tensioned concrete elements are prefabricated in a factory and must be transported to the construction site, which limits their size.

Pre-tensioned elements may be balcony components, lintels, floor slabs, beams or foundation piles and bridge decks. Figure 2.11 shows the tension cables in pre-tensioned concrete.

2.4.5.2   Bonded Post-tensioned Concrete

Bonded post-tensioned concrete results from applying compression after pouring concrete and curing in situ. The concrete is cast around a plastic, steel, or aluminium curved duct, to follow the area where otherwise tension would occur in the concrete element. Tendons are fished through the duct and the concrete is poured. Once

FIGURE 2.12   Tensioning tendon.

the concrete has hardened, the tendons are tensioned by hydraulic jacks that react against the concrete members (Figures 2.12–2.15). When the tendons have stretched sufficiently, according to the design specifications based on Hooke's law, they are wedged in position and maintain tension after the jacks are removed, transferring pressure to the concrete. The duct is then grouted to protect the tendons from corrosion. This method is commonly used to create monolithic slabs for house construction in locations where expansive soils create problems for the typical perimeter foundation. All stresses from seasonal expansion and contraction of the underlying soil are taken into the entire tensioned slab, which supports the building without significant flexure. Post-stressing is also used in the construction of bridges, both after

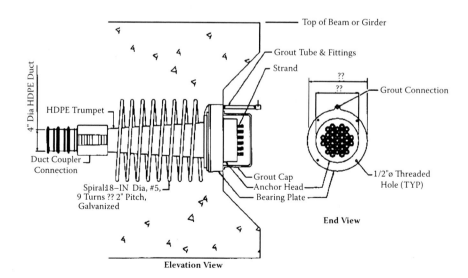

FIGURE 2.13   Elevation and end views of tendon.

FIGURE 2.14   Placing tensioning wire in bridge girder.

FIGURE 2.15   Pre-stress bridge girder.

concrete is cured after support by falsework and by the assembly of prefabricated sections, as in a segmental bridge. The advantages of this system over unbonded post-tensioning are

1. Large reduction in traditional reinforcement requirements as tendons cannot destress in accidents.
2. Tendons can be easily "woven," allowing a more efficient design approach.
3. Higher ultimate strength due to bond generated between the strand and concrete.
4. No long-term issues with maintaining the integrity of the anchor/dead end.

A standard strand is made from seven individual cold-drawn wires, six helically wound outer wires, and one center wire called a king wire. Strands can be galvanized

TABLE 2.1

Technical Data for Tendons in Different Codes and Specifications

Type	13 mm (0.6″)		15 mm (0.5″)			
Code Specifications	ASTM A416 Grade 270	prEN 10138 BS5896 Super	ASTM A416 Grade 250	prEN 10138 BS5896 Super	ASTM A416 Grade 270	prEN 10138 BS58
Yiels strength, N/mm^2	1670[a]	1580[b]	1550[b]	1500[b]	1670[a]	1580[b]
Ultimate strength, N/mm^2	1860	1960	1725	1770	1860	1860
Nominal diameter, mm	12.7	12.9	15.24	15.7	15.24	15.7
Cross sectional area, mm^2	98.71	100	139.35	150.0	140	150.0
Weight, kg/m	0.775	0.785	1.094	1.18	1.102	1.18
Ultimate load, KN	183.7	186.0	240.2	265	260.7	279
Modulus of elasticity, N/mm^2			195,000			
Relaxation[c] after 1000 h at 0.7x ultimate load			Max 2.5			

[a] Yield measured at 0.1% residual elongation.
[b] Yield measured at 1% effective elongation.
[c] Application for relaxation class 2 according to Eurocode prEN 10138/BS5896, or low relaxation complying with ASTM A416, respectively

or epoxy coated without any loss in strength including the wedge anchorage. For maximum corrosion protection a thin-walled polyethylene/polypropylene (PE/PP) plastic duct provides long-term secondary protection especially in aggressive environments such as waste-water treatment plants, acid tanks, and silos. The plastic duct is an integral part of so-called double corrosion protection tendons. Technical data for tendons is shown in Table 2.1.

2.4.5.3 Unbonded Post-tensioned Concrete

Unbonded post-tensioned concrete differs from bonded post-tensioning by providing each individual cable permanent freedom of movement relative to the concrete. To achieve this, each tendon is coated with a grease (generally lithium based) and covered by a plastic sheathing formed in an extrusion process (Figures 2.16 and 2.17). The transfer of tension to the concrete is achieved by the steel cable acting against steel anchors embedded in the perimeter of the slab. The main disadvantage over bonded post-tensioning is the fact that a cable can de-stress itself and burst out of the slab if damaged (such as during repair on the slab). The advantages of this system over bonded post-tensioning are

FIGURE 2.16   Pre-stress and pre-cast bridge girders.

FIGURE 2.17   Installing pre-stress bridge girder.

1. The ability to individually adjust cables based on poor field conditions (for example, shifting a group of four cables around an opening by placing two to either side).
2. Post-stress grouting is eliminated.
3. The ability to de-stress the tendons before attempting repair work.

2.4.6   HOLLOW SLAB

This is a structure system that relies on a concrete slab with hollows as shown in Figure 2.18 and putting steel cables in the slab through holes that will be subject to tension force as in the previous method.

There are different dimensions of the slabs, as well as each type having a specific thickness as shown in Figures 2.19–2.22. They have a thickness of 200 mm up to 400 mm. Consequently, select the thickness based on the distance between the two

FIGURE 2.18   Shape of hollow slab.

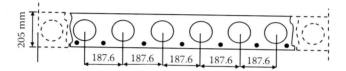

FIGURE 2.19   205 mm hollow core.

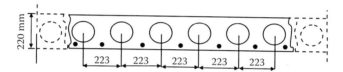

FIGURE 2.20   220 mm hollow core.

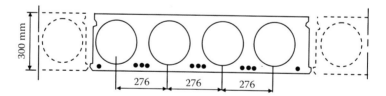

FIGURE 2.21   300 mm hollow core.

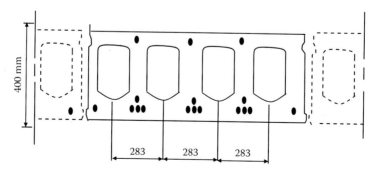

FIGURE 2.22   400 mm hollow core.

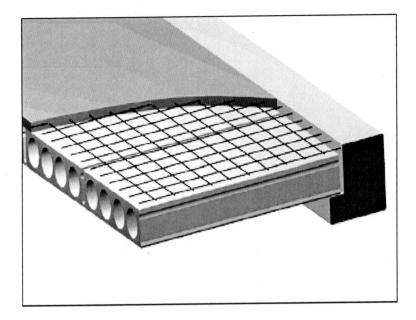

FIGURE 2.23   Pouring concrete on hollow slab.

supports as the load distribution through one direction, and in the transverse direction there is an edge with the dimensions shown in the drawings. After installing the hollow slab, light steel bars will placed on it and then concrete will be poured, as shown in Figure 2.23.

2.4.7   COMPOSITE SECTION

The composite combines a reinforced concrete section with a hot rolled steel section, taking the advantages of the two systems in maintaining the steel section's durability, as it does not need continuous coating for corrosion protection, and resisting fire (Figures 2.24 and 2.25).

Moreover, this system is usually used to reduce the size of the sections where steel strength is high in compression and tensile stress (around 140 N/mm^2) but in the case of concrete its compression strength in usual work is 25–40 N/mm^2 at 28 days. The

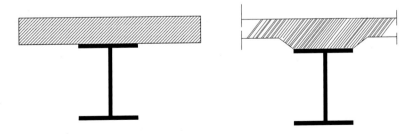

FIGURE 2.24   Two types of composite slab systems.

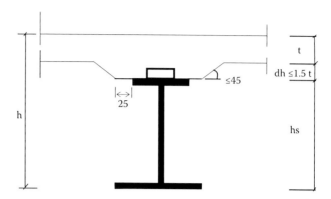

FIGURE 2.25   Haunch detail for composite slab.

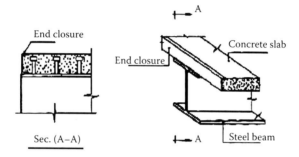

FIGURE 2.26   Relationship between concrete slab and studs.

compression design strength will be 7–10 N/mm², and in the case of tensile strength it will be around 2.5 N/mm². The steel section can carry 20 times more than concrete so the section dimensions will be reduced around 20 times.

The method of fixing the stud will be through welding or by using a stud that has special materials in its bottom (Figure 2.26). By using a special machine at high temperature the special materials will melt and then weld the stud of the steel floor, as shown in Figures 2.27 and 2.28.

The stud is a very important element to create a bond between the concrete slab and the steel. Therefore, it has specified dimensions that should be considered. Figure 2.29 shows the required dimensions. In some cases, the channel section can be used rather than the stud and it welds to the steel floor. Figure 2.30 illustrates the process of pouring concrete on the roof after installation of the studs as well as reinforcing steel bars.

A composite section beam consists of an I-beam with stirrups as shown in Figure 2.31, based on the Egyptian specifications as a guide. Concrete is then poured to fully surround the steel beam.

Composite columns have three traditional shapes as shown in Figure 2.32. In most cases, an I-beam is surrounded by stirrups and then concrete is poured to completely surround the I-beam. The other two types consist of rectangular or circular steel and are filled with concrete.

FIGURE 2.27 Fixing stud to floor steel sheet.

FIGURE 2.28 Fixation of stud.

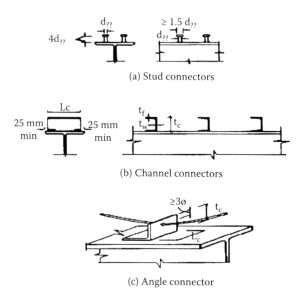

(a) Stud connectors

(b) Channel connectors

(c) Angle connector

FIGURE 2.29   Different types of connectors.

FIGURE 2.30   Pouring concrete after fixing studs.

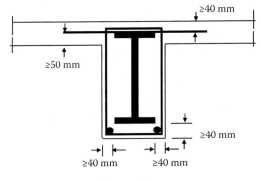

Encased Beam

FIGURE 2.31   Composite beam.

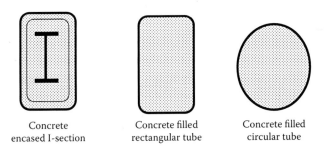

Concrete   Concrete filled   Concrete filled
encased I-section   rectangular tube   circular tube

FIGURE 2.32   Types of composite columns.

2.5   HIGH-RISE BUILDING

The high-rise building needs different structure systems to resist wind and earthquake loads. In areas with high seismic activity, the structur should have more than one structure system to resist the earthquake load. These structure systems have been developed through scientific experiments and represent practical ways to achieve adequate safety with an acceptable economic cost.

Each system can be used separately, such as shear walls or frames, or they can be used in a combined system. It is a great challenge for the structural engineer to choose the proper structure system, so in the following sections all the structure systems and how to select a suitable system are illustrated.

In the early use of reinforced concrete there were limits to building heights, but experience and research led to an increase in knowledge about the properties of concrete and the structure systems that allow taller buildings. Fazlur Khan started a revolution in the area of high-rise building from steel and concrete when he initiated the structure system for high-rise buildings. There are several systems for the structure of reinforced concrete buildings and every system has an approximate number of floors to carry from an economic point of view as shown in Figure 2.33.

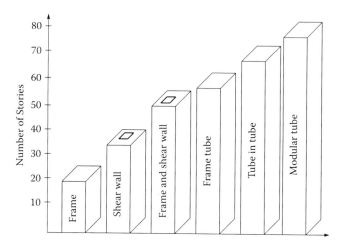

FIGURE 2.33   Different structure systems of high-rise buildings.

2.5.1 FRAMES

A frame concrete structure is one of the systems used in high-rise buildings. It is worth mentioning that the connection between the column and beam is rigid to carry the flexural momentum. Recently, using new software it has become easy to analyze and design this structure system. On the other hand, the very critical issue is constructing this rigid connection from a steel detailing point of view.

We are lucky that the design is very simple, especially with the use of the software available on the market. What is important in this system is the construction, especially in the connection between column and beam to avoid honeycomb as the column steel bars will be continuous inside the beam.

One can see that the system consists of beam and columns and force is transferred from slabs to beams and then to columns. The beam is designed to carry a shear force and bending momentum at its end so the beam and column section will be significantly bigger. Figure 2.34 presents the structure deformation due to horizontal load.

Figure 2.35 illustrates the distribution of the steel bars between beams and columns in the case of a frame structure system, and from this one can imagine

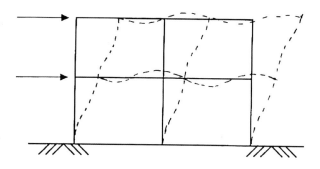

FIGURE 2.34   Deformation of frame system.

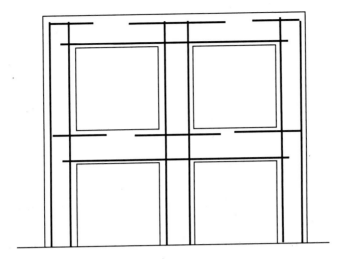

FIGURE 2.35 Generic distribution of steel bars.

the complexity of the steel detailing in the connection, which is the key in frame structures.

2.5.2 SHEAR WALL

The use of shear walls started in 1940, and they are vertical walls made from reinforced concrete that act as cantilevers. They act vertically in the form of separate planner walls or non-planner assemblies of connected walls around elevators. The shear walls transfer the horizontal loads from slabs to the foundation and increase the inertia of the building.

There are several forms of shear walls. The most traditional form is a reinforced concrete box in the core of the building, which contains the elevators or stairs, walls, or storage area.

Also, shear walls can be used internally to carry the vertical gravity loads plus the horizontal load due to wind and earthquake. There is an advantage of the shear walls as they absorb sound and also act as fire insulators between rooms and apartments, so they are usually used in hotels and residential buildings.

The shear wall system is stiffer than the frame system so it can be practical for 35–40 stories. In contrast to rigid frames, the shear wall is a solid form that tends to restrict architectural planning when open internal spaces are required.

2.5.3 SHEAR WALL AND FRAME SYSTEM

This system has shear walls with frames structure and through a study by Fazlur Khan became a main step in the development of the high-rise building.

This structure system is a combination of shear walls inside the building that uses the beam or slabs with the columns as a frame system to provide high rigidity to the beams or slabs.

This combination system is reasonable for buildings between 40 and 60 stories.

2.5.4 Framed Tube Structure

The framed tube structure system was developed by Khan in the early 1960s. The definition is a structure system of three-dimensions consisting of three, four, or more frames or shear walls connected at their ends to resist the lateral load from any direction and acts as a cantilever and transfers the load to the foundation.

The tube structure system provides stiffness in three dimensions between frames or inside the building to resist the overturning momentum and the tube makes the shear walls, columns, and beams work as one unit. The characteristics of the tube structure are that the external columns will be spaced near each other (2–4 m) and connected by a deep beam to create a special reinforced concrete structure system to resist the lateral load through three-dimensional construction. The external windows are around 50% of the external wall area. In the case of buildings that need big openings, such as malls or parking garages, internally there will be a rigid beam to transfer the lateral load and the tube system will be weak in this point. The tube structure system is acceptable architecturally and structurally.

FIGURE 2.36 DeWitt-Chestnut apartment building in Chicago.

FIGURE 2.37   The Ontario Center in Chicago.

In 1962, the first building using this tube system was the DeWitt-Chestnut apartment building in Chicago, as shown in Figure 2.36.

Several configurations of tubes exist: framed, braced, solid core-wall tubes, tube-in-tube, and bundled tubes. The framed or boxed tube is the one most likely associated with the initial definition. The DeWitt-Chestnut apartment building in Chicago is a framed tube.

A braced tube is a three-dimensionally braced or a trussed system. Its unique feature is that structure members have axial but little or no flexural deformation. Figure 2.37 is a photo of the Ontario Center in Chicago as an example of such a system in concrete. The John Hancock Center in Chicago, on the other hand, is a remarkable example of this system in steel. Tubular core walls can either carry full lateral load or they may interact with frames. The Brunswick building (Figure 2.38) in Chicago is an example where the core walls interact with the exterior frame, comprising closely spaced columns. This gives the building a tube-in-tube appearance, although it was designed using the shear wall–frame interaction principle. Tube-in-tube is a system with a framed tube and an external and internal shear wall core, which act together in resisting lateral loads. One Shell Plaza (Figure 2.39) in Houston is a tube-in-tube building. Bundled tubes are used in very large structures as a way of decreasing the surface area for wind resistance and creating intimate

FIGURE 2.38   Brunswick Building in Chicago.

spaces for occupants. Multiple tubes share internal and adjoining columns, depending on their adjacencies. One Peachtree Center in Atlanta is an example of a concrete bundled tube design. Similarly, One Magnificent Mile in Chicago is another example as shown in Figure 2.40.

Choosing a structural system is very complex in today's market. In the early years of commercial construction, only post-and-beam construction existed. Concrete's formwork was complex but putting the building together was not as complex as today's systems demand. Developments in the world of concrete since 1960 have been mostly in new systems such as tube and composite construction. The challenge for engineers and architects today is to make all the systems work together to their maximum capacity and create a habitable environment.

The Marina City Towers were built in 1962 in the center of an industrial park. Architect Bertrand Goldberg knew that whoever lived there would need 24-hour services, entertainment, parking, and offices all in one structure. Marina City contains a movie theater, bowling alley, shops, offices, restaurants, meeting rooms, gym, skating rink, parking for cars and boats and, finally, apartments. Originally marketed to single adults and couples without children, the apartment complex was a success.

The towers were two of the first new mixed-use structures in downtown Chicago and were the tallest reinforced concrete buildings in the world for that year at 588 feet (179 m). Goldberg's plan in building a circular structure, which was very innovative

FIGURE 2.39   One Shell Plaza in Houston.   **FIGURE 2.40**   One Magnificent Mile in Chicago.

at the time, was based on the efficiency of HVAC systems and reducing the service core of the structure. Another innovation was the 20-story parking garage directly below the 900 apartments. He introduced a circular core wall in hopes that it would take the entire lateral load from cantilevering floors. This was actually modified with two rows of columns so that the depth of the cantilever and beams between "petals" could be reduced. Even with these modifications, the circular core area carries 70% of the total lateral loads. The core, which acts as a circular concrete shear wall, was carefully designed with staggered openings and by minimizing their size in an effort to maintain enough stiffness.

2.5.5   Cases of High-Rise Buildings

The Water Tower Place as shown in Figure 2.41 is yet another high-rise structure of concrete located in the downtown area of Chicago and was designed in 1975 by Loebl, Schlossman, Dart & Hackl. It stands 262 m in height and serves as another mixed-use building with a mall on the interior and offices and apartments above. The strength of concrete used in this building took a dramatic jump to 62.1 MPa. This was, however, only one of 11 mixes placed by 6 cranes. Mix strengths could vary from 20.7 MPa for slabs to 62.1 MPa for columns. This building demonstrates

FIGURE 2.41   Water Tower Place.

concrete technology's ability to rival steel for tall buildings as it was two-thirds the height of the tallest steel building of the time. The structural system for Water Tower combines "reinforced concrete peripheral framed tube, interior steel columns, and a steel slab system with a composite concrete topping."

Three Eleven South Wacker Drive, built in 1990, is a super tall reinforced concrete building in Chicago (Figure 2.42). It stands at 295 m with 82.7 m as its highest concrete strength. The structural system is a modified tube with a reinforced concrete peripheral frame, interior steel columns, and a composite steel and concrete slab.

Three Eleven South Wacker Drive is a good example of shear wall–frame interaction systems. The building is engineered in such a way that the relative stiffnesses of both internal and external elements remain the same for the entire height of the building. Three strengths of concrete were used: 10,000 and 68.9 and 82.7 MPa.

A self-climbing pump with a separate, mounted, placing boom delivered concrete to the top of the structure. Post-tensioned floor slabs reduced the amount of steel while concrete consumption was reduced with thinner elements due to the material's high strength. Two sets of flying forms were cycled every five days through the structure.

One Peachtree Center, built in 1991 in Atlanta, Georgia, is a 62-story, 257 m tall, bundled-tube design with three strengths of concrete used in its columns and shear walls: 58.6, 68.9, and 82.7 MPa. Designers used technology developed in the 1960s

FIGURE 2.42   311 South Wacker Drive.

and 1970s by Material Service Corporation in Chicago. Architectural requirements dictated a column-free interior and thus a 15.2 m floor span was accomplished with HSC and post-tensioned steel bars. Silica fume and granite aggregates were used to achieve the necessary strengths. Each floor has about 36 rentable corner offices. This building is remarkable because of its design allowing multiple spaces for renters, the structural details, and its use of high-strength concrete.

Two buildings constructed around the turn of the 20th century where concrete was primarily used as the structural material in conjunction with steel are the Petronas Towers shown in Figure 2.43 in Kuala Lumpur, Malaysia (currently the tallest building in the world) and the Jin Mao building as shown in Figure 2.44 in Shanghai, China. These are good examples to show that concrete has greatly advanced as a material in only a century. The structural frames for the 452 m tall Petronas Towers use columns, core and ring beams of HSC, and floor beams and decking of steel to provide cost effectiveness, fast construction, and future adaptation to the internal and external environment. The core and frame together provide adequate lateral stiffness for such a tall building. The recently built 421 m-high Jin Mao building of 1999 is a mixed system that has a number of steel outrigger trusses tying the building's concrete core to its exterior composite mega-columns.

A summary of the different structure systems used to resist the lateral load is shown in Figure 2.45.

FIGURE 2.43   Petronas Towers.

FIGURE 2.44   Jin Mao Building.

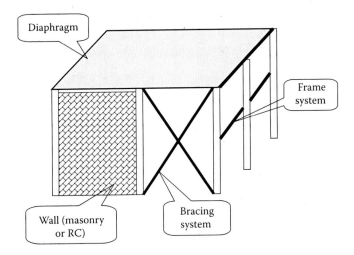

FIGURE 2.45   Different lateral force structure systems.

2.6   FOUNDATIONS

The foundation type depends on the nature of the building structure, the number of stories, loads, and the soil type in the building location. The first step is to perform the soil boring and analyze the soil with field and laboratory tests to define in general the ability of the soil to carry the load in a safe manner. Generally, there are two types of foundations: shallow and deep and these two types will be discussed.

2.6.1   SHALLOW FOUNDATION

These foundations are called rafts as they are close to the ground surface at a distance of about not less than 1.5 m and are often not more than 4 m in the case of a basement. This type of foundation is often used in the case of strong and cohesive soil, as well as when the number of building stories is few.

2.6.1.1   Isolated Footing

This type of foundation is easier to design and construct and more common as it is a less costly type of foundation. Figure 2.46 shows the shapes of the isolated footing.

2.6.1.2   Combined Footing

This is usually used in the case of adjacent columns and if it is designed as isolated columns will be merged so it needs a different design (Figure 2.47). The disadvantage of this type of foundation is more steel is necessary than in the case of the isolated footing. Therefore, the good engineering design is to avoid this expensive type of footing, especially in a small building.

2.6.1.3   Strap Footing

This type of foundation involves connecting two isolated footings by a rigid beam to carry the distributed load that affects it. This type of footing is usually used in a

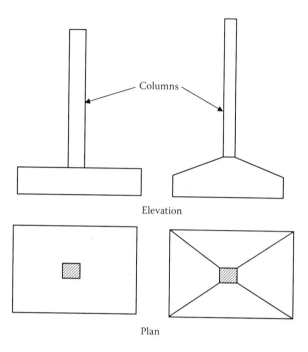

FIGURE 2.46   Isolated footing.

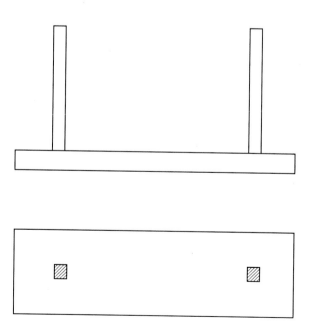

FIGURE 2.47   Combined footing.

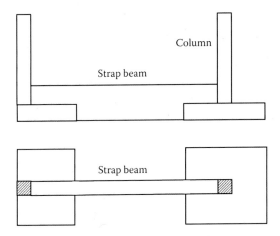

FIGURE 2.48 Strap footing.

column near a neighbor footing as shown in Figure 2.48. Because the column cannot be in the middle of the footing, overturning will occur, so this rigid beam will prevent occurrences of overturning.

2.6.2 RAFT FOUNDATION

Raft foundation is usually used in multi-story buildings, where the size of the shallow footings will be increased and combined with each other so it is essential to choose the raft foundation option (Figure 2.49).

In the past there were different methods used to design the raft foundation, but they were not realistic and did not present the soil footing interaction in a good manner. Nowadays, by using a computer, the solution of the raft foundation by using finite element methods is realistic.

It should be noted, however, that even with the most sophisticated techniques of raft analysis, the actual stresses may be different from the calculated values. The main reason is the difference among the soil behaviors assumed in the analysis. For example, the linear elastic continuum model, which may be considered as the most realistic model in raft analysis, is based on the assumption that the soil is homogeneous, elastic, and isotropic. Also, it does not differentiate between the behavior of clayey and sandy soils.

The calculated momentums and shears in the raft should be treated with caution. A generous amount of reinforcement should be provided to account for any variation from the calculated values. But in the case of a high-rise building reinforcement is a function of strong and moderate soil, but if the soil is too weak to carry the load by raft foundation, it is essential to use the deep foundation.

2.6.3 DEEP FOUNDATION

Pile foundation is the most common type of deep foundation used to transmit the structural loads into the deeper layers of firm soil in such a way that these layers of

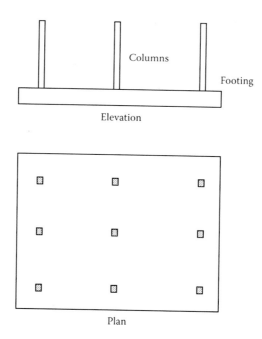

FIGURE 2.49   Raft foundation.

soil or rock can sustain the loads. A pile foundation, in general, is more expensive than an ordinary shallow foundation, and is used where soil at or near the surface is of poor bearing capacity or settlement problems are anticipated. The main functions of piles are

1. To carry more load from the superstructure to the lower, more resistant soil strata, thus increasing the load capacity of the site.
2. To reduce the settlements to the minimum value and consequently the differential settlements. They are most effective in the case of sensitive structures, which by virtue of their sensitive structural statical systems cannot undergo appreciable differential settlements.
3. To avoid excavation under water for sites where ground water table (GWT) is high. This may represent an expensive item in the cost of foundation and may also cause reduction of strength of some certain soils.
4. To increase the density of the soil by driving compaction piles in loose cohesion-free soil deposits.

Generally, piles are made of timber, steel, concrete, or a combination of these materials. Technically, there are two types of piles: end bearing piles and friction piles. The two options have advantage and disadvantages.

Figure 2.50 presents the end bearing pile; most of the load transfers to the rock layer by the end bearing of the pile. The heaviest load will be taken by its base and the rest of the load is taken by the section friction. The skin friction along the system

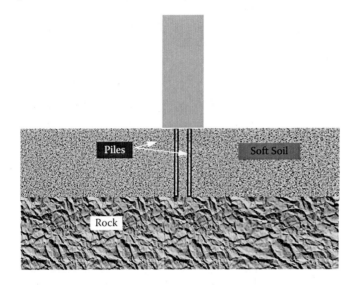

FIGURE 2.50 End bearing pile.

of the pile could be neglected and the bearing capacity of the pile is derived only from the point bearing resistance of the soil under the pile tip.

In order to obtain the full benefit of the ultimate strength of the firm layer under the pile tip, the pile should penetrate the bearing stratum to a depth at least three times the pile diameter. Figure 2.51 presents the friction piles that transfer their loads to the surrounding soil by friction developed along their sides.

If the pile penetrates a clay layer, the skin friction is equal to the cohesion, C, of that layer. In the case of granular material, the skin friction is proportional to the intensity of earth pressure and can vary linearly with enough accuracy.

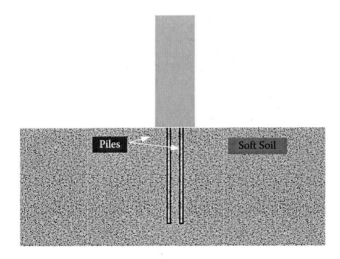

FIGURE 2.51 Friction pile.

FIGURE 2.52   Preparing wood piles.

2.6.3.1   Timber Piles

Timber piles are widely used in wooded countries and are made of tree trunks of good quality and size (not less than 300 mm in diameter) as shown in Figures 2.52 and 2.53. After the timber pile is driven into the ground, the top end should be cut off square, so that the foundation is in contact with soil wood. If a timber pile will be subjected to alternate wetting and drying, it should be treated with a wood preservative to increase its useful life. Timber piles can safely carry loads between 15 and 25 tons for usual conditions. They are of relatively low cost and easy to handle.

2.6.3.2   Steel Piles

These are usually rolled sections of H shapes or steel pipes. Wide-flange and I-beams may also be used. Splices in steel piles are made in the same manner as for steel

FIGURE 2.53   Wood pile with steel cap.

FIGURE 2.54   Steel pile with cone tip.

columns, i.e., by welding, riveting, or bolting. Pipe piles are welded or made of seamless steel which may be driven either open-ended or closed-ended.

2.6.3.3   Concrete Piles

These types of piles consist of two types: the cast-in-place piles and the precast concrete piles.

2.6.3.3.1   Cast-in-Place Piles

Cast-in-place piles are formed by making a hole in the ground and filling it with concrete. They may be drilled or formed by driving a shell. The steel piles, shown in Figures 2.54 and 2.55, are either driven or used to make a hole and then fill concrete

FIGURE 2.55   Raymond shell piles.

FIGURE 2.56   Lifting of concrete pile.

inside it. The steel shell is usually withdrawn during or after pouring concrete and sometimes is left to protect the concrete from mixing with the mud or to prevent the cement from erosion by ground water.

2.6.3.3.2   Precast Piles

Piles are formed to the specified length, cured, and then shipped to the construction site. A primary consideration is the handling stress. To take care of handling stresses, some of which are tensile, the piles are reinforced and in some cases pre-stressed.

Precast reinforced concrete (RC) piles may have square or octagonal cross sections. They should be adequately reinforced to withstand driving and handling stresses. Long precast piles should be driven with careful guides to prevent their buckling during driving in the part of the pile in the driving rig. To overcome the driving stresses the lateral steel reinforcements should be closely spaced at the top and bottom of the pile to resist the stress wave concentration at the ends.

To avoid bending momentum due to handling for relatively short piles (less than 12.0 m), the pile is usually lifted from one end and treated as a simple beam carrying its own weight. Long piles should be lifted from two, three, or four points at the specified distances indicated in Figure 2.56 to reduce the bending stresses to a minimum. Lifting points should be marked using hooks or bolts that will be removed later.

Special care should be given to the material of concrete piles to keep them intact and prevent aggressive soil from attacking the pile (Figures 2.57 and 2.58). It is necessary to use siliceous aggregates and rich cement content (350 kg of cement per m³ of finished concrete). In addition, concrete piles must be protected from the dissolved sulfates or chemicals present in the underground water. If the ratio of sulfur trioxide (SO_3) in the soil water increases over 0.03% (300 mg/l) in stagnant water, or 0.015 in running water, and if in addition the ratio of SO_3 in the soil is 0.2%, ordinary Portland cement cannot be used. In this case, special sulfate resisting cements should be used. In all cement used, the presence of free lime or calcium traces should be minimized.

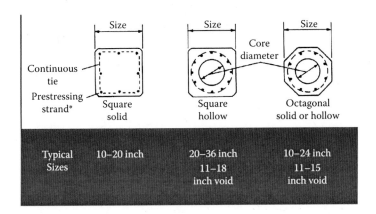

FIGURE 2.57 Typical sizes and shapes for reinforced concrete piles.

FIGURE 2.58 Composite section piles.

Table 2.2 summarizes the differences between different types of piles.

2.6.3.4 Pile Caps

Pile cap foundations transfer the column loads from the structure to the piles at the point of contact between the piles and the caps (Figure 2.59). In this type of foundation we can ignore the impact of the soils where the soils are not in contact with the caps in a rigid or flexible manner to allow them to carry any part of the column load; because the stiffness of the piles is great, they carry all the load.

Often the column load cannot be carried by only a single pile and multiple piles may be needed to carry the load. It needs a pile cap to distribute the column load equally.

In design phase, care should be taken to distribute the load to the piles equally by joining the center of gravity of the column with the center of gravity of the pile cap.

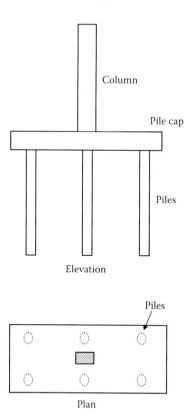

FIGURE 2.59   Sketch for pile cap.

To ensure transfer of the load from the column to the pile, the steel reinforcement should extend inside the pile cap at least 600 mm to guarantee transference of the load through the bond between the concrete and steel.

The pile caps will be designed as a rigid foundation and the piles carry equal loads from the column load so the pile cap thickness should resist the punching stresses and the tension at top and bottom.

TABLE 2.2
Typical Pile Characteristics and Uses

	Pile Type			
Characteristic	Concrete-Filled Steel Pipe Piles	Composite Piles	Precast Concrete (Including Prestress)	Cast in Place (Thin Shell Driven with Mandrel)
Maximum length	Practically unlimited	55 m	30 m for precast, 60 m for prestress	30 m for straight section, 12 m for tapered section
Optimum length	12–36 m	18–36 m	12–15 m precast; 18–30 m prestress	12–18 m for straight; 5–12 m for tapered
Application material specifications	ASTM A36 for core ASTM A252 for pipe ACI 318 for concrete	ASTM A36 for core ASTM A252 for pipe ACI 318 for concrete ASTM D25 for timber	ASTM A15 for reinforcing steel ASTM A82 for cold drawn wire ACI 318 for concrete	ACI
Recommended maximum stress	0.4 fy reinforcement, 0.5 fy or core 0.33 fc' for concrete	Same	0.33 fc' 0.4 fy for reinforcement unless prestress	0.4 fy if steel gauge ≤14; 0.35 fy if shell thickness ≤3 mm
Maximum load for usual condition	1800 KN with cores 18,000 KN for large section with steel core	1800 KN	8500 KN for prestress 900 KN for precast	675 KN
Optimum load range	700–1100 KN with cores	250–550 KN	350–3500 KN	250–725 KN

REFERENCES

Abouelei, A. 1990. *Lecture on foundations*. Cairo University.

Ali, M. M. 2001. *The art of skyscraper: genius of Fazlur Khan*. New York: Rizzoli International Publications, Inc.

Ali, M. M., and P. J. Armstrong, eds. 1995. *Architecture of tall buildings*. Council on Tall Buildings and Urban Habitat. Monograph 30. New York: McGraw-Hill, Inc.

Colaco, J. P. 1986. The mile high dream. *Civil Engineering*, ASCE. April 1986:76–78.

Egyptian code of practice. 2003. EPC203.

El Nimeiri, M. M., and F. R. Khan. 1983. *Structural systems for multi-use high-rise buildings, developments in tall buildings*. New York: Van Nostrand Reinhold Company, 221.

Elzner, A. O. The first concrete skyscraper. *The Architectural Record*. June:515, 1904.5.

Harries, K. A. 1995. Reinforced concrete at the turn of the century. *Concrete International*. January:58–62.

Huxtable, A. L. 1957. Reinforced concrete construction: the work of Ernest L. Ransome 1884–1911. *Progressive Architecture*. 38:138–42.

Khan, F. R. 1972. Future of high rise structures. *Progressive Architecture*. October.

Khan, F. R. 1972. Influence of design criteria on selection of structural systems for tall buildings. *Canadian Structural Engineering Conference*. Montreal, Canada. March:1–15.

Khan, F. R., and J. Rankine. 1980. *Structural systems, tall building systems and concepts*. Council on Tall Buildings and Urban Habitat, American Society of Civil Engineers, 42.

Khan, F. R., and J. A. Sbarounis. 1964. Interaction of shear walls and frames in concrete structures under lateral loads. *Journal of the American Society of Civil Engineers*. 90(ST3).

Malinwski, R., and Y. Garfinkel. 1991. Prehistory of concrete: concrete slabs uncovered at Neolithic archaeological site in southern Galilee. *Concrete International*. March:62–68.

Smith B. S., and A. Coull. 1991. *Tall building structures: analysis and design*. New York: John Wiley & Sons.

3 Loads in International Codes

3.1 INTRODUCTION

The philosophy for design and construction of any building is that it must have the capability to carry all the expected loads. The building's safety depends on the load fluctuations and the probability that the load will increase beyond what the building was designed for.

The types of loads and their values vary according to the building location; for example, in European countries the ice load is the most critical load that affects the design and is much different from loads in Middle East countries with very hot climates, especially in summer, so the increase in temperature is a critical factor in design.

To define the loads we must answer one very important question. What is the building design lifetime? The answer is easy in the case of residential buildings, but for commercial, industrial, and special buildings the answer depends on the project economy as a whole, and this will be clearly illustrated in the last chapter.

To know the structure's lifetime is very important in the case of loads that correlate with the probability distribution over time such as wind and earthquake loads, and in the case of offshore structures the wave load also.

3.2 LOADS

The main steps included in building design are to define the load on the structure and calculate structure resistance to these loads, so defining the load value is the first step in the structure design process. The load values are defined according to the codes and specifications that the designer will consider. There are different types of loads as will be shown in the following sections.

3.2.1 DEAD LOAD

The dead load is the structure's self weight, which can be defined by knowing the structure's dimensions and the density of the element materials such as concrete, steel, wood, etc. So the first step is defining the preliminary member dimensions. In addition, we must know the material from which the member will be constructed.

The dead load includes the weight of finishing such as plastering, tiles, and walls that are constructed from bricks and other finishing materials that are not considered the main parts. It is very important to know that the dead load has a very high value from the total load during design; therefore, the weight of the building members must be calculated precisely, especially in the case of a reinforced concrete structure. The densities of different materials usually used in building are shown in Table 3.1, to be considered in calculating the dead load.

Recently, computer software has been used in structure analysis to calculate the self-weight of the structure member based on defining its materials and dimensions. However, the finishing materials and the dead load that occupy the building should be input into the program.

3.2.2   Live Load Characteristics

In the last decade, there have been significant advances of the application of structure reliability theory (El-Reedy et al. 2000). One area of attention has been live loads in buildings, including the development of realistic stochastic models. Rapid advances of probabilistic modeling of live loads in recent years can be attributed to a growing awareness of the designer's uncertainty about the loads acting on a structure and the acceptance of a probabilistic assessment of these loads.

The total live load is composed of two parts, which are the sustained load and the extraordinary load. The summation of these two types of load is presented to obtain the probabilistic model of the total live load on a residential building along the lifetime of the structure.

The time behavior of the live load model on a given floor area can, in general, be divided into two parts: a sustained load and an extraordinary (or transient) load. The sustained load includes the furnishings and occupants normally found in buildings and is usually measured in live load surveys. This load is assumed to be a spatially varying random function and constant with time until a load change takes place. These load changes are assumed to occur as Poisson arrivals as shown in Figure 3.1a. The extraordinary load is usually associated with special events that lead to high concentrations of people, and it may also be due to the stacking of furniture or other items. The extraordinary load occurs essentially instantaneously and it is assumed to arrive as a Poisson event. Each event is modeled by a random number of load values as shown in Figure 3.1b.

The total live load history shown in Figure 3.1c is the sum of sustained and extraordinary load components, and its maximum value represents the largest total load that may be incurred on a given floor area during the structural lifetime.

3.2.2.1   Stochastic Live Load Models

Most live loads acting on structures are of random nature. In addition many such loads fluctuate with time. Determining the total combination of such loads requires a solution within the scope of the theory of random processes wherein loads are described as stochastic processes.

According to Renold (1989), a stochastic process is a collection of random variables when the collection is infinite. The collection will be indexed in some way, e.g., by calling $\{x(t):t \in T\}$ a stochastic process, we mean that $x(t)$ is a random variable for each (t) belonging to some index set (T). (More generally, $x(t)$ may be a vector of random variables.)

TABLE 3.1
Self-Weights of Different Materials

Material Type	Kg/m³
Plain concrete	2200
Reinforced concrete	2500
Lightweight concrete	1000–2000
Air-entrained concrete	600–900
Heavy concrete	2500–5500
Cement (loose)	1200–1100
Clinker	1800–1500
Big aggregate	1700
Sand	1500
Foamed aerated slag	1700
Granulated aggregate	1200
Expanded clay	900–200
Pumice stone	650–350
Exfoliated vermiculite	200-60
Fly ash	1100–600
Water	1000
Liquid or powder	1200–1000

Masonry Stones

Igneous Rocks

Granite	2800
Basalt	3000
Basalt lava	2400
Trechzte	2600

Sedimentary Rocks

Limestone	2700
Marble	2800
Sandstone	2700

Transformed Rocks

Slate	2800
Gneiss	3000
Serpentine	2700
Marble	2700

Masonry Bricks

Red brick	1800–1600
Solid brick	1850
Hollow brick	1400
Light brick	800–700

Continued

TABLE 3.1 (*Continued*)
Self-Weights of Different Materials

Material Type	Kg/m³
Refractory Brick for General Purposes	
Fire clay	1850
Silica	1800
Magnisite	2800
Chrome-magnisite	3000
Corundum	2600
Brick, antiacid	1900
Glass brick	870
Masonry Block	
Concrete block	1400–1900
Hollow concrete block	1150
Leca concrete block	600–800
Gypsum block, lime	950
Lime	
Limestone powder	1300
Calcined in lumps	850–1300
Calcined	600–1300
Calcined and slaked	110
Gypsum	800–1000
Mortar	
Cement mortar	2100
Lime mortar	1800
Lime cement mortar	750–1800
Gypsum mortar	1400–1800
Bitumen mortar with sand	1700
Wood Types	
Hard Wood	
Beech	680
Oak	790
Soft Wood	
Pitch pine	570
White wood	400
Fiber Board	
Hard	900–1100
Medium-hard	600–850
Porous insulating	250–400

TABLE 3.1 (*Continued*)
Self-Weights of Different Materials

Material Type	Kg/m³
Plywood	750–850
Core board	450–650

Other Building Materials

Asbestos	800
Asbestos board	1600
Asbestos cement pipe	1800
Celton	120
Dry earth	1700
Wet earth	2000
Rubber floor	1800
Asphalt	3200
Bitumen	1000–14000
Tar	1100–1400
Cement tile	2400
Mosaic tile	2200

Epoxy Resin

Without fill	1150
With mineral material	2000
With fiberglass	1800
Plastic tile	1100
Polyester resin	1350
Polyetherene	930
PVC hard board	1400
PVC flooring board	1600
PVC flooring tile	1700
Fiberglass	160–180
Glass wool	100–110
Slag wool	200–300
Cork	60
Plaster	1100–1500
Glass in sheets	2500
Wired glass	2600
Acrylic glass	1200
Linen, baled	600
Leather in piles	1000–900

Paper

Stocks	1200
Rolls	1100

Continued

TABLE 3.1 (Continued)
Self-Weights of Different Materials

Material Type	Kg/m³
Rubber	
Rolled flooring	1300
Raw, balled	1100
Wool	
Bales	700
Pressed and baled	1300
Metallic Materials	
Steel	7850
Wrought iron	7850
Cast iron	7250
Iron ore	3000
Aluminum	2700
Aluminum alloy	2800
Lead	
White lead (powder)	9000
Red lead (powder)	8000
Copper	890–8700
Brass	8500–8300
Bronze	8900
Nickel	7900
Zinc	8200
Zinc, rolled	7200
Tin, rolled	7400–7200
Magnesium	1850
Antimony	6620
Barium	3500
Cadmium	8650
Cobalt	8700
Gold	19,300
Silver	10,500
Manganese	7200
Molybdenum	10,200
Platinum	21,300
Titanium	4500
Tungsten	19,000
Uranium	5700
Vanadium	5600
Zirconium	6530

TABLE 3.1 (Continued)
Self-Weights of Different Materials

Material Type	Kg/m³
Fuels	
Mineral coal	1200–900
Coke	650–450
Charcoal	250
Coal dust	700
Oils	
Diesel oil	1000–800
Crude oil	950
Petrol (gasoline)	800–850
Petroleum	800
Liquid Gas	
Propane	500
Butane	580
Wood	
Hard wood, chopped	600–400
Hard wood, logs	500
Soft wood, chopped	250
Soft wood, logs	300
Fire wood	400
Liquids	
Glycerine	1250
Oil paint, canned or boxed	1100
Milk	
Tanks	1000–950
Box	1000
Cans	600
Honey	
Tanks	1300
Cans	1000
Bottles	600
Acids	
Nitric acid	1500
Hydrochloric acid	1200
Sulfuric acid	1400

Continued

TABLE 3.1 (Continued)
Self-Weights of Different Materials

Material Type	Kg/m³
Foodstuffs and Agricultural Products	
Butter in	
Barrels	550
Boxes	800–500
Sugar in	
Paper	600
Big boxes	800
Lump sugar in paper sacks	600
Boxed	700
Other	
Tea packets	400
Alcohol	800
Beer in tanks	1000
Beer in barrels	900
Cacao in boxes	550
Eggs	550
Fat, boxed	800
Fish in barrels	600
Fish in cans	800
Fruits in boxes	400–350
Fruits	700–500
Hay, baled	200–150
Maize corn	450
Margarine in barrels	550
Margarine in boxes	700
Meat, refrigerated	700–400
Onions in bags	550
Pickles in sacks	700
Drinks, bottled	800
Rice	500
Rice in bags	560
Salt in piles	1,000
Salt in bags	1120
Starch flour in bags	800
Straw, baled	170
Tobacco, baled	500–300
Wheat	900–800
Wine in tanks	1000
Wine in barrels	850

TABLE 3.1 (*Continued*)
Self-Weights of Different Materials

Material Type	Kg/m³
Coffee in bags	700
Soap powder in sacks	610

Other Materials

Books and files	1100–1000
Ice in blocks	900–850
Textiles	1100
Cellulose, baled	800
Cloth, baled	400
Cloth, baled	1300–700
Cotton, baled	500
Hemp, baled	400
Jute, baled	700

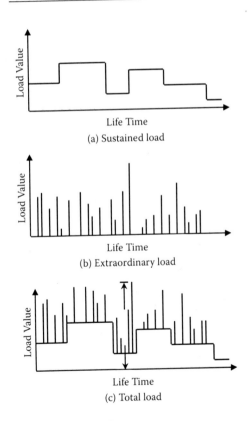

FIGURE 3.1   Live load characteristics.

Usually the index is either an interval $T = [0, \infty]$, or a countable set such as the set of nonnegative integers $t = \{0, 1, \ldots\}$, and t will denote time with these alternatives of T. We have either a continuous or discrete time process, respectively.

Starting with Borges and Castanheta's work (1972), a number of models for describing the variability of loads have been suggested by Bosshard (1975), Grigoriu (1975), Larrabee (1978), Pier and Cornell (1973), and Wen (1977). Among these models, the most general yet tractable models are the Poisson square wave (PSW) and the filter Poisson (FP) processes.

3.2.2.2   Poisson Square Wave Process

This process has been used to model sustained loads. It assumes that the load intensity changes at random points in time following a Poisson process. Intensities within these points are assumed to be independent and identically distributed random variables.

If (v_s) denotes the mean rate of load changes, and F_S (s) describes the common cumulative density function (CDF) of intensities, Parzen (1967) shows that the occurrences of $S > s$ following load changes constitute a Poisson process with mean occurrence $v_s [1 - F_S (s)]$. The CDF of the extreme may, therefore, be evaluated as

$$F_{S_T} = P(S_T < s) = P(S < s)_{t=0} P(noS > s.following.load.changes)_{0 \to t}$$
$$= F_S(s)\exp\{-v_S T[1 - F_S(s)]\}$$

(3.1)

Upon differentiating, the probability density function (PDF) is obtained as

$$f_{S_T}(s) = f_S(s) [1 + v_s T F_s(s)] \exp\{-v_s T [1 - F_S(s)]\}$$

(3.2)

in which $f_S(s)$ is the PDF of S, and T is the total lifetime.

3.2.2.3   Filtered Poisson Process

This process has been used to model extraordinary loads with random occurrences. It is assumed that load occurrence follows a Poisson process, and that load intensities at various occurrences are independent and identically distributed random variables. The process is generalized to include random load duration, denoted by t, which is also assumed to be statistically independent and identically distributed at various occurrences.

Let v_r denote the mean occurrence rate of the extraordinary load, $F_R(r)$ represent the distribution at each occurrence, and let (N) be the random number of occurrences during (T). On the basis of the total probability theorem, the CDF of R_T is obtained as

$$F_{RT} = P(R_T < r) = \sum_{n=0}^{\infty} P(R < r/N = n) P(N = n)$$

(3.3)

in which

$$P(N = n) = \frac{(v_r T)^n \, Exp(-v_r T)}{n!} \tag{3.4}$$

and

$$P(R < r/N = n) = [F_r(r)]^n \quad \text{for } n = 1, 2, 3, \ldots \tag{3.5}$$

$$P(R < r/N = n) = 0.0 \qquad R < 0.0; \text{ for } n = 0$$

$$P(R < r/N = n) = 1.0 \qquad R > 0.0; \text{ for } n = 0$$

The last equation neglects the effect of overlapping of occurrences and is therefore acceptable when overlapping is unlikely or when load intensities at overlaps do not accumulate. By substitution of equations (3.4) and (3.5) in equation (3.3), one can obtain the following:

$$F_{R_T}(r) = \sum_{n=0}^{\infty} [F_R(r)]^n \frac{(v_r T)^n \, Exp(-v_r T)}{n!}$$

$$F_{R_T}(r) = [F_R(r)]^0 \, Exp(-v_r T) + \sum_{n=1}^{\infty} [F_R(r)]^n \frac{(v_r T)^n \, Exp(-v_r T)}{n!}$$

$$= [F_R(r)]^0 \, Exp(-v_r T) + F_R(r) v_r T . Exp(-v_r T) + [F_R(r)]^2 \frac{(v_r T)^2}{2!} \, Exp(-v_r T) + \ldots$$

$$F_{R_T}(r) = Exp\{-v_r T[1 - F_R(r)]\} + (H_r(r) - 1) Exp(-v_r T) \tag{3.6}$$

in which H(r) is unit step function

$$H_r(r) = 0 \qquad \text{if } r < 0.0 \tag{3.7}$$

$$H_r(r) = 1.0 \qquad \text{if } r > 0.0$$

Upon differentiating equation (3.6), the PDF of R_T is obtained as

$$f_{R_T} = v_r \cdot T \cdot f_R(r) \cdot exp[-v_r T \cdot [1 - f_R(r)] + \delta(r) \cdot exp(-v_r T) \tag{3.8}$$

in which $\delta(r)$ is the Dirac delta function representing a spike unit area at $r = 0.0$.

3.2.2.4 Analysis of Suggested Model

The total live load is the summation of sustained load and extraordinary load over the lifetime of the structure.

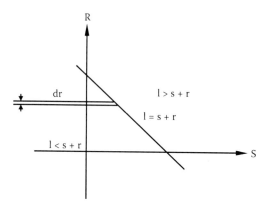

FIGURE 3.2   Region of summation for two random variables not restricted to nonnegative values.

To describe the sum of the sustained and extraordinary random load variables, consider the independent continuous random variables, s and r, with probability density function $f_S(s)$ and $f_R(r)$, and cumulative distribution functions $F_S(s)$ and $F_R(r)$, for sustained and extraordinary load, respectively. Assuming independence, the joint probability density function $f_{S,R}(s,r) = f_S(s) \cdot f_R(r)$ (Cornell and Benjamin 1974). To determine the PDF of the total load, l, where, l is linear function of s and r

$$l = s + r$$

We first find CDF of l where the joint probability density function of s and r is integrated over the region $s + r \leq l$ as indicated in Figure 3.2:

$$F_L(l) = P(L \leq l) = P(l \leq s + r)$$

$$= \int_{-\infty}^{\infty} \int_{-\infty}^{l-r} f_{S,R}(s,r)dsdr$$

$$= \int_{-\infty}^{\infty} \int_{-\infty}^{l-r} f_S(s) \cdot f_R(r)dsdr \tag{3.9}$$

$$= \int_{-\infty}^{\infty} F_S(l-r)f_R(r)dr$$

Then, the corresponding PDF can be obtained by differentiating the CDF of Z as follows:

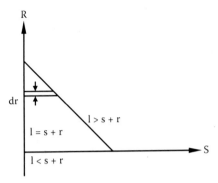

FIGURE 3.3 Region of summation for two random variables restricted to nonnegative values.

$$f_L(l) = \frac{\partial F_L(l)}{\partial L}$$

$$= \frac{\partial}{\partial L} \int_{-\infty}^{l-r} F_S(l-r) \cdot f_R(r) dr \qquad (3.10)$$

$$= \int_{-\infty}^{\infty} f_S(l-r) \cdot f_R(r) dr$$

where the density f_L is called the convolution of probability densities $f_S(s)$ and $f_R(r)$. For nonnegative random variables as in the case of live loads, the range of integration in equations (3.9) and (3.10) shown in Figure 3.2 reduces to the region indicated in Figure 3.3 because $f_S(s)$ and $f_R(r)$ are zero for negative argument; therefore, equations (3.9) and (3.10) become

$$F_L(l) = \int_0^l F_S(l-r) \cdot f_R(r) . dr \qquad (3.11)$$

$$f_L(l) = \int_0^l f_S(l-r) \cdot f_R(r) dr \qquad (3.12)$$

Substituting equations (3.1), (3.2), (3.6), and (3.8) for the PDF and the CDF for sustained and extraordinary loads, respectively, into equations (3.11) and (3.12) to obtain the PDF and CDF of the total load is as follows:

$$f_L(l) = \int_0^1 f_s(s)\big[1 + v_s T f_s(s)\big] Exp\{-v_s T[1 - f_s(s)]\}$$
$$\cdot v_r T \cdot f_R (1 - s) Exp\{-v_s T[1 - f_s(1 - s)]\} ds \tag{3.13}$$

$$F_L(l) = \int_0^1 Exp\{-v_r T[1 - F_R(1 - s)]\} f_s(s)\big[1 + v_s T F_s(s)\big]$$
$$\cdot Exp\{-v_r T[1 - F_R(s)]\} ds \tag{3.14}$$

where

T is the lifetime of the structure.

v_s is the mean rate of load change in case of sustained load.

v_r is the mean occurrence rate of the extraordinary.

$F_s(s)$ and $f_s(s)$ are the cumulative density function and the probability density function of the sustained load, respectively.

$F_R(r)$ and $f_R(r)$ are the cumulative density function and the probability density function of the extraordinary load, respectively.

Knowing the structure lifetime, the duration of the sustained load, the rate of occurrence of extraordinary load, and the probability density function and cumulative density function of the sustained and extraordinary loads, and performing the integration in equations (3.13) and (3.14), one can find the values of the probability density function, $f_L(l)$, and cumulative density function, $F_L(l)$, corresponding to the value of the total load l.

Several researchers discussed these parameters and their values for different types of building occupation. Chalk and Corotis (1980) summarized survey data results for sustained and extraordinary load. They found that the gamma distribution is presented by the sustained load value and the exponential distribution is used to represent the time between changes. On the other hand, the extraordinary load value is presented by a gamma distribution and its occurrence as a Poisson distribution.

After reviewing survey data, Corotis and Doshi (1977) substantiated the use of gamma probability distribution of the magnitude of sustained load. On the other side, Cornell and McGuire (1974) suggested using gamma probability distribution to present the probability density function of a single extraordinary event.

In ANSI code, the statistical parameters for sustained and extraordinary load and the duration of sustained load and rate of occurrence of extraordinary load are presented for different occupation types as shown in Table 3.1. Moreover, the reference periods of different occupancies are presented.

According to Michael et al. (1981), all extraordinary live load occupancy durations are assumed to follow a uniform distribution with mean value 2 weeks, 6 hours, and 15 minutes. Der Kiureghiam (1980) takes the duration of the extraordinary load equal to 3×10^{-3} year.

According to Der Kiureghian (1978) and (1980), the extraordinary load durations are short enough with respect to mean sustained load durations that the

TABLE 3.2

ANSI Code (A58.1-1982): Typical Live Load Statistics

| Occupancy | Sustained Load | | Extraordinary Load | | Temporal Constants | | |
	m_s (kg/m²)	σ_s (kg/m²)	m_r (kg/m²)	σ_r (kg/m²)	τ_s (years)	v_e (per year)	T (years)
			Office Buildings				
Offices	53.2	28.8	39.1	40.0	8	1	50
			Residential				
Owner occupied	29.3	12.7	29.3	32.2	2	1	50
Renter occupied	29.3	12.7	29.3	32.2	10	1	50
			Hotels				
Guest rooms	22.0	5.9	29.3	28.3	5	20	50
			Schools				
Classrooms	58.6	13.2	33.7	16.6	1	1	100

extraordinary loads may still be considered point processes when combined with the sustained loads for lifetime total statistics. The sustained load which occurs during the structure lifetime (T) only is of interest considering the sustained load interval of Poisson square wave process between zero and the first transition point and noting that a portion of the last interval is generally truncated. Therefore, the mean rate of sustained load duration will be modified and calculated by the following equations:

$$v_{ok} \cong v_r\, v_s\, (\tau_r + \tau_s)$$

$$v_{sm} \cong \left(\frac{0.5}{T} + v_s \right) \frac{v_{ok}}{v_s + v_{ok}} \tag{3.15}$$

where
v_{ok} is the mean rate of coincidence in this combination.
τ_r is the duration of the extraordinary load.
τ_s is the duration of the sustained load.
v_{sm} is the modified sustained load, mean rate.

Table 3.2 from the American National Standards Institute (ANSI) presents the average values for live loads and standard deviations for different uses of buildings for sustained and extraordinary loads.

3.2.2.5 Methodology and Calculation Procedure

The integration of equations (3.13) and (3.14) by the analytical procedure is very complicated and time consuming. The Romberge numerical integration technique, which is described in many mathematical books, is used in this section to obtain the value of the total live load.

The Romberge numerical integration technique is performed by using a software program called Excel (1996). The probability density function of the total live load is calculated assuming that the parameters of equations (3.13) and (3.14), which are used in the integration, are as follows:

- The structure lifetime (T) is 50 years.
- The sustained live load is assumed to have a gamma distribution with mean value kg/m² with standard deviation 16.6 kg/m² and mean occurrence rate, v_s, is 0.5/year.
- The extraordinary load is assumed to have a gamma distribution with mean value equal to 29.29 kg/m² and standard deviation of 32.22 kg/m² and the mean occurrence rate of 1.0/year.

The duration time of extraordinary load, τ_r is 3×10^{-3} year. Therefore, the modified sustained load mean occurrence rate is calculated from equation (3.15) and found to be 0.34/year.

The probability density function of total live load is obtained as plotted. The mean and standard deviation of the total load are calculated and the mean total load in the case of a residential building is equal to 106.13 kg/m² and the standard deviation is equal to 42.52 kg/m².

3.2.2.6 Testing of Suggested Model

After the model distribution is obtained by the previous integration method, the data is tested with the common distributions by using chi square (χ^2), which is goodness-of-fit applied to continuous random variables and its value is calculated from the following equation:

$$\chi^2 = \sum_{i=1}^{k}\left[\frac{(O_i - E_i)^2}{E_i}\right] \tag{3.16}$$

where O_i and E_i are the observed and expected number of occurrences in the ith interval, respectively, and k is the number of intervals.

Moreover, the Kolmograv-Smirnov (K-S) test is done, which is a second quantitative goodness-of-fit test based on a second test statistic. It concentrates on the deviations between the hypothesized cumulative distribution function and the observed cumulative data, and the calculation is based on the following equation:

$$K - S = \max_{i=1}^{n}\left[\left|\frac{i}{n} - F_X(X^{(i)})\right|\right] \tag{3.17}$$

The results of the two tests with some hypothesized probability distributions are shown in Table 3.3, and the lowest value for the two tests is at hypothesized

TABLE 3.3
Goodness-of-Fit Results Comparing Suggested Model and Common Probability Distributions

Probability Distribution	Chi-χ^2	K-S
Lognormal distribution	0.004644	0.01226
Extreme value type I	0.019781	0.02022
Gamma distribution	0.04796	0.036673

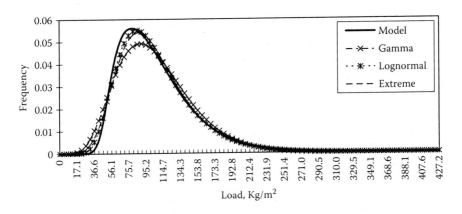

FIGURE 3.4 Comparison of model and different probabilities distributions.

lognormal distribution. Besides that, the relations between the suggested model and the different probability distributions are shown in Figure 3.4, which makes it obvious that this model coincides more with the lognormal distributions than the others.

Moreover, the test by the above two methods may be performed by the Crystal Ball program (1996) to test the suggested model with the more traditional probability distributions. Note that the model is more near to the lognormal distribution with mean value equal to 106.13 kg/m² and standard deviation equal to 42.52 kg/m².

3.2.2.7 Verification of Suggested Model Using Monte-Carlo Simulation

The research was performed by El-Reedy et al. using the Monte-Carlo simulation technique for constructing a model for a residential building, and this research revealed that lognormal distribution is the best probability distribution that can present the live load for a residential building. The comparison of the model and different distributions are shown in Figure 3.4.

In order to verify the analytical model with the suggested load of occupancy, a Monte-Carlo simulation technique (Ang and Tang 1984) was used to simulate the total load process by using the Crystal Ball program (1996) as follows:

1. Generate the magnitude of the sustained load from the gamma distribution.
2. Generate the duration of the sustained load at the previous magnitude from the exponential distribution.
3. From the previous duration time and by using Poisson distribution generate the number of the extraordinary load occurrences during the sustained load duration.
4. Generate the extraordinary load magnitude for every load from extraordinary load gamma distribution.
5. Calculate the total load value by summing the sustain load and every extraordinary load.
6. Repeat steps 1 to 5.

The Monte-Carlo simulation is performed for 10,000 trials. The values from the trials are divided into 5 kg/m² intervals. The frequency of occurrences at each interval is plotted as a bar and the corresponding values from the suggested model as a curve as shown in Figure 3.5. There is no difference between the suggested analytical load model and Monte-Carlo simulation results.

The analysis shows that the live load effect along the lifetime of the residential building, taking into consideration the lifetime of the structure years, can be presented by a lognormal distribution with mean value equal to 106.13 kg/m² and standard deviation equal to 42.52 kg/m².

Rapid advances of probabilistic modeling of live loads in recent years can be attributed to a growing awareness of the uncertainty about the loads acting on a structure and the acceptance of a probabilistic assessment of these loads.

As normal procedure, from the probability distribution, obtain the definite number that can be used easily by the design engineer as the equations of design for any code are based on the deterministic analysis. But the risk and probability of failure differ from one code to another so the values of the loads are different.

3.2.2.8   Live Loads in Different Codes

The values of a live load and its reduction factors are different from one code to another, depending on the country that is applying the code. Moreover, the resistance

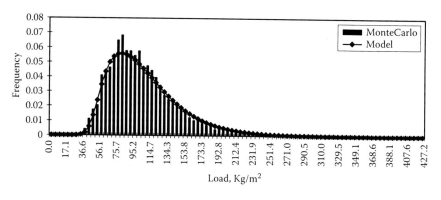

FIGURE 3.5   Comparison of model and Monte-Carlo simulation results.

reduction factor is different from one code to another depending on the quality control, the capabilities of the contractors, and the government system of each country. Therefore, the reliability of a reinforced concrete structure is different from one code to another depending on factors of safety of load and resistance.

In the following section, a comparison of live load values for different codes (ECP, BS8110, ANSI) is presented.

3.2.2.8.1 Comparison of Live Loads for Different Codes

Live load factor accounts for the unavoidable deviations of the actual load from the code value and for uncertainties in the analysis that transforms the load into a load effect.

The load reduction factor is recommended by different codes to account for the decrease of total load due to the reduced probability of applying the same value of load at different floors at the same time. The values and factor of live load are different from one code to another, while the reduction factor of live load is different from one floor to another in residential buildings under different design codes.

Therefore, in the following sections, comparisons between codes of design are presented to show the difference between the values of live load, limit state load factors, and reduction of floor load in three codes. The codes that are considered in this comparison are

- Egyptian code of practice (ECP)
- British code (BS8110)
- American National Standards Institute (ANSI) code

3.2.2.8.2 Values of Live Load and Its Factors In Different Codes

The live load values that are taken into consideration in design are different from one code to another depending on the country. The different live load values for different codes are shown in Table 3.4. The Egyptian code is more conservative for the value of live load. However, the factor of live load is the same in limit state equations for most of the codes, but ANSI code is more conservative than the others. It is very important to illustrate the methods of calculating live load in Egyptian, American code (ANSI), and the British code (BS8110).

TABLE 3.4
Comparison of Live Load Factors
under Different Codes

Code	Load value (kg/m²)	Equation
ECP	200	1.4 D.L + 1.6 L.L
BS8110	153	1.4 D.L + 1.6 L.L
ANSI	195	1.4 D.L + 1.7 L.L

The live load values that are considered in design will be based on building use. All the values for the different codes are illustrated in Table 3.5.

3.2.2.8.3   Floor Load Reduction Factor in Different Codes

The reduction factors of live load in different floors in residential buildings vary in the codes considered in this section (ECP, BS8110, and ANSI codes). These reduction factors for live load are summarized in Table 3.6. It is worth mentioning that the reduction factor in ANSI code depends on the influence area. Members having an influence area of 37.2 m² (400 ft²) or more may be designed for a reduced live load determined by applying the following equation:

$$L = Lo \cdot \left(0.25 + \frac{15}{\sqrt{A_I}} \right)$$

(3.18)

where

L = reduced design life load per square foot of area supported by the member.

Lo = unreduced design live load per square foot of area supported by the member.

A_I = influence area, in square feet.

The influence area A_I is four times the tributary area for a column. For instance, in the case of an interior column, the influence area is the total area of the floor surrounding bays.

The total load carried by columns at a certain floor level is calculated by summing the load reduction factors, given in Table 3.6, for the floors above this level as a function of load values given in Table 3.5 for the considered codes. The results are given in Table 3.7 as a function of P_1, P_2, and P_3, which are different live load values in ECP, BS8110, and ANSI.

Moreover, the total reduction of live load carried by columns at different floors varies from the ECP to the BS8110 and ANSI codes. These values in ECP are greater than those in the BS8110 and ANSI. However, these values are the same in the ECP and BS8110 from the 10th to the 20th floors.

Generally, the ANSI code provides lower values for the total live load reduction factor than other codes.

3.2.2.8.4   Comparison of Total Design Live Load Values under Different Codes

After comparing the reduction factors and live load values in different codes, the total design live load of a column carrying a certain number of typical floors is compared. The total design live load of that column is calculated by summing the floor loads, and the sum is multiplied by the corresponding reduction factor specified by the codes. The ratio between the total live load value calculated from ECP and that from other codes is obtained, as presented Table 3.8.

From this table, the total live load value taken in design according to ECP is more conservative than the values obtained from the other codes. The total live load calculated using the ECP is higher than that obtained from BS8110 and ANSI, but the

TABLE 3.5
Live Load Values under Different Codes

| | | | Live Load | |
| | | | BS8110 | |
Type of Building	ECP (kg/m²)	ANSI[a] (kg/m²)	Distributed Load (kg/m²)	Concentrated Load (kg)
1. Roof				
Horizontal surface cannot reach (no use)	100			
Inclined >20° (no use)	50			
Horizontal or inclined (can be reached and used) in residential building	200			
Horizontal (can reach in public buildings)	300			
2. Residential building				
Rooms	200	195.3	153	143
Stairs	300	195.3	306	459
Balcony	300	195.3	153	153/m'
3. Administration building	300			
Halls		488	408	459
Offices		244	255	275
4. Theaters, cinemas, and libraries		500	488	408
5. Conference room without fixed chairs		600		408
6. Warehouse	1000			
Light		610.3	408/m height storage	918
Heavy		1220.61	765	459
7. School				
Classrooms		195.3	306	275
Corridors		390.6	408	459
Stairs and exits		488	408	459
8. Hospital				
Operation room and laboratory		293	204	459
Diagnosis room		195.3		
Reception hall		195.3	204	459
Corridor above first floor		390.6	408	459

[a] ANSI is the American National Standards Institute.

TABLE 3.6
Floor Reduction Factors for Live Load in Different Codes

Floor	Floor Reduction Factor		
	ECP	BS8110	ANSI[a]
Roof	1.0	1.0	0.82
1st	1.0	0.9	0.65
2nd	0.9	0.8	0.58
3th	0.8	0.7	0.54
4th	0.7	0.6	0.51
5th	0.6	0.6	0.48
6th	0.5	0.6	0.47
7th	0.5	0.6	0.45
8th	0.5	0.6	0.44
9th	0.5	0.6	0.43
10th	0.5	0.6	0.42
11th	0.5	0.5	0.41
12th	0.5	0.5	0.41
13th	0.5	0.5	0.41
14th	0.5	0.5	0.41
15th	0.5	0.5	0.41
16th	0.5	0.5	0.41
17th	0.5	0.5	0.41
18th	0.5	0.5	0.41
19th	0.5	0.5	0.41

[a] Calculated from equation (3.18) for four bays of 16 m².

TABLE 3.7
Comparison of Reduction Factors under Different Codes

No. of Floors	ECP	BS8110	ANSI[a]
6	$5P_1$	$4.6P_2$	$3.58P_3$
8	$6P_1$	$5.8P_2$	$4.5P_3$
10	$7P_1$	$7P_2$	$5.4P_3$
12	$8P_1$	$8P_2$	$6.2P_3$
14	$9P_1$	$9P_2$	$7P_3$
16	$10P_1$	$10P_2$	$7.8P_3$
18	$11P_1$	$11P_2$	$8.6P_3$
20	$12P_1$	$12P_2$	$9.4P_3$

Note: P_1, P_2, and P_3 are the different live load values given in Table 3.4.

[a] Calculated from equation (3.18) for four bays of 16 m².

TABLE 3.8

Comparison of Design Live Loads for Egyptian Code, BS, and ANSI

Floor No.	EC/BS8110	EC/ANSI
1	1.31	1.17
6	1.42	1.35
8	1.35	1.29
10	1.31	1.26
12	1.31	1.24
14	1.31	1.24
16	1.31	1.23
18	1.31	1.23
20	1.31	1.23

ratio between values of the total live load calculated from ECP to values obtained from ANSI and BS8110 is different from one floor to another.

In the case of six floors, the Egyptian code has a more conservative design live load value by about 42% than the British code. Moreover, ECP live load design value is higher than BS8110 for 10 to 20 floors by about 31%. On the other hand, the ratio between the live load calculated from ECP to that from ANSI code has a minimum value equal to 1.17 at one floor only, but the highest ratio is 1.35 at six floors with average values of 1.25.

In summary, the Egyptian code in calculating live load in the case of the limit state design method is more conservative than other codes of design.

When making a comparison of different specifications for a building of 20 stories, the parameters (P_1), (P_2), and (P_3) are the values of live load in the specifications.

The Egyptian code is more conservative followed by the British code and then the American code, and these factors are affected by customs and traditions.

For example, in the case of developed countries, the rules for changing the activity of a building from residential to administration or to industrial are fewer. Builders should be more conservative in design parameters than in countries that have stricter laws for changing a building's activities.

Table 3.6 compares the Egyptian, American and British codes according to the number of floors. Generally, the Egyptian code is more conservative than British code by about 30%, and the table gives the highest values in the American code by about 23%. These differences are fixed in the case of 16 floors and more, and for fewer numbers of floors the values are as shown in Table 3.7.

3.2.3 WIND LOAD

Wind load governs the design based on the building's geographic location and the shape of the building. Therefore, the parameters that govern the wind load are approximately similar in the different codes as all are based on aerodynamic theory.

TABLE 3.9

ANSI Code Classification of Buildings and Other Structures for Wind and Earthquake Loads

Category	Nature of Occupancy
I	All buildings and structures except those listed below
II	Buildings and structures where the primary occupancy is one in which more than 300 people congregate in one area
III	Buildings and structures designated as essential facilities, including, but not limited to: (1) Hospital and other medical facilities having surgery or emergency treatment areas (2) Fire or rescue and police stations (3) Primary communication facilities and disaster operation centers (4) Power stations and other utilities required in an emergency (5) Structures having critical national defense capabilities
IV	Buildings and structures that represent a low hazard to human life in the event of failure, such as agricultural buildings, certain temporary facilities, and minor storage facilities

3.2.3.1 ANSI Code

Every building or structure and every portion thereof shall be designed and constructed to resist the wind effects determined in accordance with the requirements of this code (Table 3.9). Wind shall be assumed to come from any horizontal direction. No reduction in wind pressure shall be taken because of the shielding effect of adjacent structures.

Structures sensitive to dynamic effects, such as buildings with a height-to-width ratio greater than five; structures sensitive to wind-excited oscillations, such as vortex shedding or icing; and buildings over 121.9 m in height shall need a special dynamic structure analysis or a wind tunnel test.

This code does not apply to building and foundation systems in those areas subject to scour and water pressure by wind and wave action. Buildings and foundations subject to such loads shall be designed according to special design precautions.

The main equation to calculate wind load is

$$q_z = 0.00256 \, K_z \, (IV)^2 \tag{3.19}$$

where

V is the basic wind speed based on the building location to the country map.
I is the importance factor.
KZ is the velocity pressure exposure coefficient.

3.2.3.2 Wind Tunnel

English military engineer and mathematician Benjamin Robins (1707–1751) invented a whirling arm apparatus to determine drag and performed some of the first experiments in aviation theory.

Sir George Cayley (1773–1857), the "father of aerodynamics," also used a whirling arm to measure the drag and lift of various airfoils. His whirling arm was 5 feet long and attained top speeds between 10 and 20 feet per second. Armed with test data from the arm, Cayley built a small glider that is believed to have been the first successful heavier-than-air vehicle to carry a man.

However, the whirling arm does not produce a reliable flow of air impacting the test shape at a normal incidence. Centrifugal forces and the fact that the object is moving in its own wake mean that detailed examination of the airflow is difficult. Francis Herbert Wenham (1824–1908), a council member of the Aeronautical Society of Great Britain, addressed these issues by inventing, designing, and operating the first enclosed wind tunnel in 1871.

Figure 3.6 presents the model of the planned largest building in the world, which is under construction in Dubai. This model prototype is put in the wind tunnel to perform the test. To study the effect of wind on the building it is important to include the surrounding buildings and to also build prototype models for the buildings around the building under study. This is shown very clearly in Figure 3.7.

The wind tunnel helps in carrying out studies on the wind loading, aeroelastic stability, and dynamic response of ground-based structures such as tall buildings, towers, and bridges. These studies are often performed in advance of construction in order that the design engineers can obtain accurate information with respect to the structural loads and deflections arising from the static and dynamic action of wind.

FIGURE 3.6 Prototype of highest building in the world in Dubai.

FIGURE 3.7 Prototype models for the buildings surrounding the building being studied.

Wind tunnel testing also assists investigations on existing structures in order to alleviate unacceptable levels of vibration. Modifications to the aerodynamic shape of the structure are then incorporated. Typically, these investigations are carried out on scale models immersed in a correctly scaled dynamic representation of the natural wind.

Moreover, wind tunnel testing measures the pressure at certain points of structures and structures are designed based on these measurements. Usually with very tall buildings or a building with an unusual or complicated shape (like a parabolic or hyperbolic shaped tall building) or cable suspension bridges or cable stayed bridges, wind tunnel testing provides the necessary design pressures for use in the dynamic analysis of the structure.

In other instances, prototype structures or elements of structures are tested at full-scale in the wind tunnels. Examples include stay cables for cable-supported bridges, overhead power lines, communications antennas, road signs, and wind turbines.

3.2.3.3 Wind Load in British Specifications

The calculation of wind load in the British standard is very precise and every factor is described in detail. It is important to note that the wind load on a partially completed structure will be dependent on the method and sequence of construction and this situation may be critical. It is reasonable to assume that the maximum design wind speed Vs will not occur during a construction period and a reduced factor ($S3$) can be used to calculate the probable maximum wind. The assessment of wind load should be made as follows:

1. *The basic wind speed V* appropriate to the district where the structure is to be erected is determined in accordance with its location on a wind map. V is the 3-second gust speed estimated to be exceeded on the average

once in 50 years. This speed has been assessed for the United Kingdom by statistical analysis of the continuous wind records from the meteorological stations after adjusting them, as necessary, to a common basis. The values are given as isopleths (lines of equal wind speed) drawn at 2 m/s intervals on the map. Values from this map represent the 3-second gust speed at 10 m (33 ft) above ground in an open situation that is likely to be exceeded only once in 50 years. V is greatest from the direction of the prevailing winds, that is, from southwest to west in the United Kingdom. For buildings, whose design may be directionally dependent, a reduced wind speed may be used for other directions, at the designer's discretion.

2. The basic wind speed is multiplied by factors $S1$, $S2$, and $S3$ to give *the design wind speed Vs.*

$$Vs = V\ S1\ S2\ S3 \qquad (3.20)$$

3. The design wind speed is converted to *dynamic pressure q* using the relationship

$$q = kVs$$

4. The dynamic pressure q is then multiplied by an appropriate pressure coefficient Cp to give the *pressure p exerted at any point on the surface* of a building.

$$P = Cpq \qquad (3.21)$$

If the value of the pressure coefficient Cp is negative this indicates that p is suction as distinct from a positive pressure.

Since the resultant load on an element depends on the difference of pressure between opposing faces, pressure coefficients may be given for external surfaces Cpe and internal surfaces Cpi. The resultant wind load on an element of surface acts in a direction normal to that surface and then

$$F = (Cpe - Cpi)qA \qquad (3.22)$$

where A is the area of the surface.

A negative value for F indicates that the resultant force is outwards. The *total wind load* on a structure may be obtained by vectorial summation of the loads on all the surfaces.

3.2.3.3.1 Topography Factor S1

V represents the wind speed as a contour line, and takes into account of the general level of the site above sea level. It is worth mentioning that it does not allow for local topographic features such as hills, valleys, cliffs, escarpments, or ridges, which can significantly affect the wind speed in their vicinity. Near the summits of hills or the crests of cliffs, escarpments, or ridges the wind is accelerated. Where the average slope of the ground does not exceed 0.05 within a kilometer radius of the site, the terrain may be taken as level and the topography factor $S1$ should be taken as 1.0.

In the vicinity of local topographic features, $S1$ is a function of the upwind slope and the position of the site relative to the summit or crest, and will be within the range of $1.0 \leq S1 \leq 1.36$. $S1$ will vary with height above ground level, at a maximum near to the ground, and reducing to 1.0 at higher levels.

In certain steep-sided, enclosed valleys, wind speeds may be less than in level terrain. Caution is necessary in applying $S1$ values less than 1.0 and specialist advice should be sought in such situations.

3.2.3.3.2   Ground Roughness, Building Size, and Height above Ground, Factor S2

Factor $S2$ takes account of the combined effect of ground roughness, the variation of wind speed with height above ground, and the size of the building or component part under consideration.

In conditions of strong wind, the wind speed usually increases with height above ground. The rate of increase depends on ground roughness and also on whether short gusts or mean wind speeds are considered. This is related to building size to take account of the fact that small buildings and elements of a building are more affected by short gusts than are larger buildings, for which a longer wind-averaging period is more appropriate.

3.2.3.3.2.1   Ground Roughness   For code purposes, ground roughness is divided into four categories and buildings and their elements are divided into three classes, as cited in 3.2.3.3.2.2 below.

Ground roughness 1. Long fetches of open, level, or nearly level country with no shelter. Examples are flat coastal fringes, fens, airfields, and grassland or farmland without hedges or walls around the fields.

Ground roughness 2. Flat or undulating country with obstructions such as hedges or walls around fields, scattered windbreaks of trees, and occasional buildings.

Ground roughness 3. Surfaces covered by numerous large obstructions. Examples are well-wooded parkland and forest areas, towns and their suburbs, and the outskirts of large cities. The general level of rooftops and obstructions is assumed at about 10 m, but the category will include built-up areas generally apart from those that qualify for category 4.

Ground roughness 4. Surfaces covered by numerous large obstructions with a general roof height of about 25 m or more. This category covers only the centers of large towns and cities.

3.2.3.3.2.2   Cladding and Building Size   Wind speed fluctuates from moment to moment and can be averaged over any chosen period. The shortest time scale, 3 s, that is normally measured produces gusts whose dimensions envelop obstacles up to

20 m across. The longer the averaging time, the greater is the linear length encompassed by the gust. For this reason three classes have been selected.

Class A. All units of cladding, glazing, and roofing and their immediate fixings and individual members of unclad structures.

Class B. All buildings and structures where neither the greatest horizontal dimension nor the greatest vertical dimension exceeds 50 m.

Class C. All buildings and structures whose greatest horizontal dimension or greatest vertical dimension exceeds 50 m.

The values of $S2$ for variation of wind speed with height above ground for the various ground roughness categories and the building size classes are given in Table 3.10.

The height should measure to the top of the structure or, alternatively, the height of the structure may be divided into convenient parts and the wind load on each part calculated, using $S2$ that corresponds to the height above ground of the top of that part. Generally, the load should be applied at the mid height of the structure or part, respectively (Figure 3.8). This also applies to pitch roofs.

TABLE 3.10
Ground Roughness, Building Size, and Height above Ground, Factor $S2$

H (m)	(1) Open country with no obstructions			(2) Open country with scattered windbreaks			(3) Country with many windbreaks; small towns; outskirts of large cities			(4) Surfaces with large and frequent obstructions, as city centers		
	Class			Class			Class			Class		
	A	B	C	A	B	C	A	B	C	A	B	C
<3	0.83	0.78	0.73	0.72	0.67	0.63	0.64	0.60	0.55	0.56	0.52	0.47
5	0.88	0.83	0.78	0.79	0.74	0.70	0.70	0.65	0.60	0.60	0.55	0.50
10	1.0	1.00	0.95	0.90	0.93	0.83	0.78	0.74	0.69	0.67	0.62	0.58
15	1.03	0.99	0.94	1.00	0.95	0.91	0.88	0.83	0.78	0.74	0.69	0.64
20	1.06	1.01	0.96	1.00	0.95	0.91	0.88	0.83	0.78	0.74	0.69	0.64
30	1.09	1.05	1.00	1.03	0.98	0.94	0.95	0.90	0.85	0.79	0.75	0.70
40	1.12	1.08	1.03	1.07	1.03	0.98	1.01	0.97	0.92	0.90	0.85	0.79
50	1.14	1.10	1.06	1.10	1.06	1.01	1.05	1.01	0.96	0.97	0.93	0.89
60	1.15	1.12	1.08	1.12	1.08	1.04	1.08	1.04	1.00	1.02	0.98	0.94
80	1.18	1.15	1.11	1.14	1.10	1.06	1.10	1.06	1.02	1.05	1.02	0.98
100	1.20	1.17	1.13	1.17	1.13	1.09	1.13	1.10	1.06	1.10	1.07	1.03
120	1.22	1.19	1.15	1.19	1.16	1.12	1.16	1.12	1.09	1.13	1.10	1.07
140	1.24	1.20	1.17	1.21	1.18	1.14	1.18	1.15	1.11	1.15	1.13	1.10
160	1.25	1.22	1.19	1.22	1.19	1.16	1.20	1.17	1.13	1.17	1.15	1.12
180	1.26	1.23	1.20	1.24	1.21	1.18	1.21	1.18	1.15	1.19	1.17	1.14
200	1.27	1.24	1.21	1.25	1.22	1.19	1.23	1.20	1.17	1.20	1.19	1.16

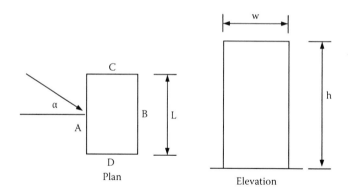

FIGURE 3.8   Building plan and elevation parameters.

3.2.3.3.3   Statistical Factor S3

Factor $S3$ is based on statistical concepts, which take account of the degree of security required and the period in years during which there will be exposure to wind.

Whatever wind speed is adopted for design purposes, there is always a probability, however small, that it may be exceeded in a storm of exceptional violence; the greater the period of years over which there will be exposure to the wind, the greater is this probability. Normally $S3 = 1$ except for:

1. Temporary structures
2. Structures for which a shorter period of exposure to the wind may be expected
3. Structures facing a longer period of exposure to the wind
4. Structures requiring greater than normal safety

For these special cases both the period of exposure to the wind and the probability level may be varied according to circumstances. In no case is the period of exposure to be shorter than 2 years.

To clarify the importance of $S3$, the wind speed having a return period, T, of 50 years should more properly be called that speed which will be exceeded just once with a probability $p = 0.02$ in any one year. Return period and probability are connected by the relation $Tp = 1$.

Now the probability that a value less than or equal to wind speed x will occur in one year is $q = 1 - p$. In a period of N years the probability Q that a value less than or equal to x will occur is the probability P that a value greater than x will occur at least once in a period of N years:

$$P = 1 - Q$$

$$= 1 - q^N \qquad (3.23)$$

$$= 1 - (1 - p)^N$$

TABLE 3.11

Cp Factor

Building Height Ratio	Building Plan Ratio	Wind Angle	Cp for Surfaces				Local Cp
			A	B	C	D	
h/w ≤ 0.5	1 < L/w ≤ 1.5	0	+0.7	−0.2	−0.5	−0.5	−0.8
		90	−0.5	−0.5	+0.7	−0.2	
	1.5 < L/w < 4	0	+0.7	−0.25	−0.6	−0.6	−1.0
		90	−0.5	−0.5	+0.7	−0.1	
0.5 <h/w ≤ 1.5	1 < L/w ≤ 1.5	0	+0.7	−0.25	−0.6	−0.6	−1.1
		90	−0.6	−0.6	+0.7	−0.25	
	1.5 < L/w <4	0	+0.7	−0.3	−0.7	−0.7	−1.1
		90	−0.5	−0.5	+0.7	−0.1	
1.5 < h/w < 6	1 < L/w ≤ 1.5	0	+0.8	−0.25	−0.8	−0.8	−1.2
		90	−0.8	−0.8	+0.8	−0.25	
	1.5 < L/w < 4	0	+0.7	−0.4	−0.7	−0.7	−1.2
		90	−0.5	−0.5	+0.8	−0.1	

For $N = 50$ and $T = 50$, $P = 0.63$. Therefore, there is only a probability of 0.63 that the once in 50 years wind speed will be exceeded at least once in 50 years.

Factor $S3$ has been obtained by selecting values of P and N, solving the equation for T, and calculating values x_T, the wind speed of return period T, for all stations in the United Kingdom where observations are available. The value of $S3$ for given P and N is $S3 = x_T/x$, where x is the once in 50 years wind speed at the same place. For each pair of values of P and N the values of $S3$ at places in the United Kingdom showed little variation with site. Therefore, the mean of $S3$ values at all places was taken for each pair of P and N.

For the calculation of wind loads during construction or for calculation of wind loads on temporary structures whose probable life is short, wind speeds may be reduced using factor $S3$. It is undesirable to consider an exposure period of less than 2 years in this context, although the critical period may be only 2 weeks. As an example, with $P = 0.63$ the value of $S3 = 0.77$ for an exposure period of two years.

Therefore, there is a probability of 0.63 that a speed which is 0.77 times the once in 50 years wind speed will be exceeded at least once in a period of two years. Because normally a probability level of 0.63 will be used for all design work, the loads during construction may be calculated using a wind speed equal to 0.77 times the map speed.

3.2.3.3.4 Dynamic Pressure of the Wind

From the value of the design wind speed, Vs obtained from the dynamic pressure of the wind q above atmospheric pressure is obtained from the Tables 3.11–3.15 according to the units used; these are derived from the equation:

$$q = kVs$$

TABLE 3.12
Wind Pressure (q)

Location	Wind Pressure (kg/m²)
Mersa Matrough	90
Alexandria, Hurgada	80
Cairo, Assuit, and other coastal areas	70
Sewa oasis	60
Fayoum, Aswan, Luxor, Tanta, Mansoura	50

Values of k are as follows for the various units used in this code:

$k = 0.613$ in SI units (N/m² and m/s).
$k = 0.062\ 5$ in metric technical units (kg/m² and m/s).

3.2.3.4 Wind Load in Egyptian Code

When performing structure analysis, wind load affects the whole structure as one unit including the foundations and the structure elements such as walls, ceilings, etc. In calculating the effect of wind on the walls and partitions and all parts of the building exposed to pressure or withdrawal of wind on the two faces, the wind carrying the design of these parts is total forced pressure or drag on the back face. The equation of calculating the wind load will be as follows for external pressure or suction:

$$Pe = CeKq \qquad (3.24)$$

where

Pe = external wind pressure.

q = the wind pressure, which depends on the building geographic location as in Table 3.13.

K = exposure factor, which changes with the height from the ground as shown in Table 3.13.

Ce = pressure distribution factor or wind external suction on the building roof, which depends on the building geometry.

1. In the case of a building that has height/width or height/length more than 2.5:

$$C = 1.3 \sin\alpha - 0.5$$

2. In the case of a building with other ratios:

$$C = 1.2 \sin\alpha - 0.5$$

TABLE 3.13
K Values

Height, Z, m	K Values
0–10	1.0
10–20	1.1
20–30	1.3
30–50	1.5
50–80	1.7
80–120	1.9
120–160	2.1
More than 160	2.3

TABLE 3.14
Internal Pressure Factors for Building with Rectangular Faces

Location of Opening	Ci
Most openings in building face in front of wind direction	+0.7
Most openings in back face	−0.5
Most openings in two faces parallel to wind direction	−0.7
Openings distributed to four faces	+0.3
Most openings in faces in front of wind direction and back face	−0.2

The equation for calculating the wind load will be as follows for external pressure or suction:

$$Pi = Ci\, K\, q \qquad (3.25)$$

where Ci is the internal factor to the internal surfaces and it depends on the location of the opening in the building views. Its values are shown in Table 3.14.

In some buildings and structures that require the calculation of the distribution of wind pressure on the roofs, especially those where the ratio of height or length to other dimensions are very high, it is preferable to calculate the total wind force on the whole structure rather than calculating distribution to unit area for this type of use, and the equation for calculating total force winds is as follows:

$$F = C_f\, K\, q\, A \qquad (3.26)$$

where

A = exposure area.
C_f = wind force factor as in Table 3.15.

TABLE 3.15
Total Wind Factor C_f

Plan		h/d	
Square shape (wind perpendicular to member)	1	7	25
Square shape (wind in diagonal dimensions)	1.2	1.3	1.3
Six or eight members	1	1.1	1.5
Circular shape with smooth surface (d'/d = 0.0)	1	1.2	1.4
Circular shape with moderately smooth surface (d'/d = 0.0)	0.5	0.6	0.7
Circular shape with rough surface (d'/d = 0.02)	0.7	0.8	0.9
Circular shape with very rough surface (d'/d = 0.08)	0.8	1	1.2

Note: d', depth of protruding element such as rips and spoilers; d, diagonal or small dimensions in plan; h, height.

In the case of special structures, as in the following conditions:

- Buildings and structures having a height over 60 meters
- Structures having a height more than four times the structure's dimensions
- Buildings and structures with unconventional shapes
- Buildings and structures in unusual areas such as on mountain peaks
- Structures with susceptibility to vibration under unusual wind such as with suspended ceilings

The following steps are required.

1. Obtain the maximum average value for the wind speed from the nearest weather record station to the building location. The measuring level from the ground should be defined and the nature of weather station location.
2. The principal wind pressure will be calculated based on the previous collecting data by statistical method to define the design wind speed.
3. Take as a guide a previous practical test on a similar structure in a wind tunnel.
4. Use the dynamic analysis technique to calculate the structure forces on the internal members.
5. For any wind load calculation, determine the effect of the wind on the building not less than the design wind load by Equation 3.26.

3.2.4 EARTHQUAKE LOADS

Earthquake is a sudden and rapid shock that takes place in the earth's crust when layers of the earth's crust slide over each other; rocks are bent to adapt to the new situation, and when the rocks return to their original position in a sudden motion the earth's surface is shaken and quake waves spread from the quake center through the earth's crust.

FIGURE 3.9 Earthquake damage.

Earthquake load is an important factor that can seriously affect a structure as shown in Figure 3.9. Earthquake loads impact buildings at intervals up to several years but inflict their damages within seconds. About 300,000 earthquakes occur annually; most are weak or occur in uninhabited areas. Those that occur in occupied areas may also cause landslides.

The United States has a number of codes that are similar in some respects but exhibit differences related to seismic activities, for example, the basic building code (BOCA), the national building code, the standard building code, and the uniform building code (UBC).

The equivalent lateral force method in UBC which is used in the western United States is more effective and widely used so it will be discussed in the following section.

3.2.4.1 UBC 1997

In most critical projects the uniform building code is usually used in calculating the earthquake load that affects the building. This code is based on ability to resist a minimum total lateral load V, which shall be assumed to act nonconcurrently in orthogonal directions parallel to the main axes of the structure.

The procedures and the limitations for the design of structures shall be determined considering seismic zoning, site characteristics, occupancy, configuration, structural system, and height in accordance with this section (Tables 3.16 and 3.17). Structures shall be designed with adequate strength to withstand the lateral

TABLE 3.16
Seismic Zone Factor

Zone	1	2A	2B	3	4
Z	0.075	0.15	0.2	0.3	0.4

TABLE 3.17
Soil Factor

Soil Profile Type	Generic Description for Soil Profile	Average Soil Properties for Top 100 Feet (30,480 mm) of Soil Profile		
		Shear Wave Velocity, vs (m/s)	Standard Penetration Test, N (or NCH for cohesionless soil layers) (blows/foot)	Undreamed Shear Strength (kpa)
Sa	Hard rock	>1500	—	—
SB	Rock	1500–760	>50	100
SC	Very dense soil and soft rock	760–360	15–50	50–100
SD	Stiff soil profile	360–180	<15	<50
SE	Soft soil profile	<180	>50	100
SF	Soil requiring site-specific evaluation			

Note: Soil profile SE includes any soil profile with more than 3 m of soft clay defined as a soil with plasticity index, PI > 20, and moisture content ≥ 45 and Su < 24 kpa.

displacements induced by the design basis ground motion, considering the inelastic response of the structure and the inherent redundancy, over strength, and ductility of the lateral-force-resisting system.

Base shear force, V, will be calculated from the following formula:

$$V = \frac{C_v I}{RT} W \tag{3.27}$$

Note that the design base shear need not exceed the following equation:

$$V = \frac{2.5 C_a I}{R} W \tag{3.28}$$

The total design base shear shall not be less than the following equation:

$$V = 0.11 C_a I W \tag{3.29}$$

In addition to that, in zone 4, the lateral force will be not less than the following equation:

$$V = \frac{0.8 Z N_V I}{R} W \tag{3.30}$$

TABLE 3.18
Seismic Coefficient, C_a

Soil Profile	Z = 0.075	Z = 0.15	Z = 0.2	Z = 0.3	Z = 0.4
SA	0.06	0.12	0.16	0.24	0.32 Na
SB	0.08	0.15	0.20	0.30	0.40 Na
SC	0.09	0.18	0.24	0.33	0.40 Na
SD	0.12	0.22	0.28	0.36	0.44 Na
SE	0.19	0.30	0.34	0.36	0.36 Na
SF	Site-specific geotechnical investigation and dynamic site response analysis shall be performed to determine seismic coefficients for soil profile type SF.				

TABLE 3.19
Seismic Coefficient, C_v

Soil Profile	Z = 0.075	Z = 0.15	Z = 0.2	Z = 0.3	Z = 0.4
SA	0.06	0.12	0.16	0.24	0.32 Na
SB	0.08	0.15	0.20	0.30	0.40 Na
SC	0.13	0.25	0.32	0.45	0.56 Na
SD	0.18	0.32	0.40	0.54	0.64 Na
SE	0.26	0.50	0.64	0.84	0.96 Na
SF	Site-specific geotechnical investigation and dynamic site response analysis shall be performed to determine seismic coefficients for soil profile type SF.				

For all buildings, the value of T can be calculated from the following equation:

$$T = C_t \left(h_n \right)^{3/4} \tag{3.31}$$

where
W is the building dead load.
C_a and C_v are obtained from Tables 3.18 and 3.19.
R depends on the structure system and its ductility.
$C_t = 0.0731$ for reinforced concrete moment-resisting frames and eccentrically braced frames.

The above equation is used for regular structures up to 73 m in height, but for irregular structures these equations can be used up to five floors with maximum 19.4 m height.
In all buildings the vibration period will be calculated from the following equation:

$$T = C_t \left(h_n \right)^{3/4} \tag{3.32}$$

where

$C_t = 0.0853$ for steel frame.

$C_t = 0.0731$ for reinforced concrete structure and frame system.

$C_t = 0.0488$ for all types of buildings.

W is the design weight for the building and it will take weight equal to the dead load (Tables 3.20–3.23).

- For a warehouse, it will be 25% from the live load.
- For a building partition, add load 0.48 KN/m^2.
- In the case of ice, the load increases to 30 Lb/ft^2; take its value into consideration and it can decrease by 75%.

3.2.4.2   Dynamic Analysis

The dynamic analysis is complicated and needs a competent engineer for a complicated structure to reduce cost the code.

Dynamic analysis is used for buildings over 73 m in height, buildings with more than five floors or height above 19.8 m in seismic zones 3 and 4, or buildings on SF soil exhibiting velocity exceeding 0.7 s.

3.2.4.3   Spectrum Analysis Method

This is calculated based on a 10% probability of increased earthquakes over 50 years. The shape of the spectrum has a relation to the parameters C_a and C_v as shown in Figure 3.10

The elastic dynamic analysis will be performed and define the dynamics for different shape modes. By using the dynamic characteristic like the natural period and natural mode, which is calculated by the modal analysis, the structure analysis is performed in three dimensions.

In the case of the regular building the lateral forces will not be less than 90% from the calculated force statically as in the previous method. However, in case of irregular structures the calculated lateral force will not be less than the value calculated from static method by about 80%.

3.2.4.4   Earthquake Calculation in ECP

The basic concept for the earthquakes loads in the Egyptian code is to set the minimum controls for the impact of earthquakes in the design of regular buildings but the code does not consider the special buildings.

The code is based on the severity of the earthquake and structure system so the building can respond to a medium intensity shake without cracks in the structure and respond to shocks of relatively high intensity without a complete collapse.

The earthquakes can be analyzed to three vectors: two horizontally and the third vertical. In conducting the structure analysis it is assumed that the two earthquake horizontal forces affect the two main directions of the building but do not affect in the same time.

It is important to know that the maximum effect of the wind load or earthquake to be taken into consideration is based on one load at one time.

TABLE 3.20
Structural Systems Configuration

Basic Structural System	Lateral-Force-Resisting System Description	R	Height Limit for Seismic Zones 3 and 4 (feet)
Bearing wall system	1. Light-framed walls with shear panels		
	a. Wood structural panel walls for structures three stories or less	5.5	65
	b. All other light-framed walls	4.5	65
	2. Shear walls		
	a. Concrete	4.5	160
	b. Masonry	4.5	160
	3. Light steel-framed bearing walls with tension-only bracing	2.8	65
	4. Braced frames where bracing carries gravity load		
	a. Steel	4.4	160
	b. Concrete[a]	2.8	
	c. Heavy timber	2.8	65
Building frame system	1. Steel eccentrically braced frame (EBF)	7.0	240
	2. Light-framed walls with shear panels		
	a. Wood structural panel walls for structures three stories or less	6.5	65
	b. All other light-framed walls	5.0	65
	3. Shear walls		
	a. Concrete	5.5	240
	b. Masonry	5.5	160
	4. Ordinary braced frames		
	a. Steel	5.6	160
	b. Concrete[a]	5.6	
	c. Heavy timber	5.6	65
	5. Special concentrically braced frames		
	a. Steel	6.4	240
Momentum-resisting frame system	1. Special momentum-resisting frame (SMRF)	8.5	N.L.
	a. Steel	8.5	N.L.
	b. Concrete[b]	6.5	160
	2. Masonry momentum-resisting wall frame (MMRWF)	5.5	
	3. Concrete intermediate momentum-resisting frame (IMRF)[c]		
	4. Ordinary momentum-resisting frame (OMRF)	4.5	160
	a. Steel[d]	3.5	
	b. Concrete[f]		
	5. Special truss momentum frames of steel (STMF)	6.5	240

TABLE 3.20 (*Continued*)
Structural Systems Configuration

Basic Structural System	Lateral-Force-Resisting System Description	R	Height Limit for Seismic Zones 3 and 4 (feet)
Dual system	1. Shear walls		
	a. Concrete with SMRF	8.5	N.L.
	b. Concrete with steel OMRF	4.2	160
	c. Concrete with concrete IMRF[c]	6.5	160
	d. Masonry with SMRF	5.5	160
	e. Masonry with steel OMRF	4.2	160
	f. Masonry with concrete IMRF[a]	4.2	
	g. Masonry with masonry MMRWF	6.0	160
	2. Steel EBF		
	a. With steel SMRF	8.5	N.L.
	b. With steel OMRF	4.2	160
	3. Ordinary braced frames		
	a. Steel with steel SMRF	6.5	N.L.
	b. Steel with steel OMRF	4.2	160
	c. Concrete with concrete SMRF[a]	6.5	
	d. Concrete with concrete IMRF[a]	4.2	N.L.
	4. Special concentrically braced frames		
	a. Steel with steel SMRF	7.5	160
	b. Steel with steel OMRF	4.2	
Cantilevered column building system	1. Cantilevered column elements	2.2	357[e]
Shear wall-frame interaction system	Concrete		160

Note: N.L., no limit.

[a] Prohibited in seismic zones 3 and 4.
[b] Includes precast concrete.
[c] Prohibited in seismic zones 3 and 4.
[d] Ordinary moment-resisting frames in seismic zone 1.
[e] Total height of building including cantilevered columns.
[f] Prohibited in seismic zones 2A, 2B, 3, and 4.

Dividing the Arab Republic of Egypt by earthquake intensity effect and excluding earthquake load for the buildings that have a height not more than 18 m from the foundation level in case of zone of weak earthquake effect, and 15 m in the second zone which is between weak to moderate, and 12 m in the third area of intensity moderate earthquake, the following conditions exist:

- Residential use.
- Concrete structures from beam and columns.
- Frame structures have reasonable rigidity in the two main directions.

TABLE 3.21
Seismic Source Type and Magnitude[a]

Seismic Source Definition[b] Source Type	Seismic Source Description	Maximum Moment Magnitude, M	Slip Rate, SR (mm/year)
A	Faults that are capable of producing large magnitude events and that have a high rate of seismic activity	$M \geq 7.0$	$SR \geq 5$
B	All faults other than types A and C	$M \geq 7.0$ $M < 7.0$ $M \geq 6.5$	$SR < 5$ $SR > 2$ $SR < 2$
C	Faults that are not capable of producing large magnitude earthquakes and that have a relatively low rate of seismic activity	$M < 6.5$	$SR \leq 2$

[a] Subduction sources shall be evaluated on a site-specific basis.
[b] Both maximum moment magnitude and slip rate conditions must be satisfied concurrently when determining the seismic source type.

TABLE 3.22
Near-Surface Factor Na1

Seismic Source Type	≤2 km	4 km	≥10 km
A	1.5	1.2	1.0
B	1.3	1.0	1.0
C	1.0	1.0	1.0

TABLE 3.23
Near-Source Factor Nv

Seismic Source Type	Closest Distance to Known Seismic Source			
	≤2 km	5 km	10 km	≥15 km
A	2.0	1.6	1.2	1.0
B	1.6	1.2	1.0	1.0
C	1.0	1.0	1.0	1.0

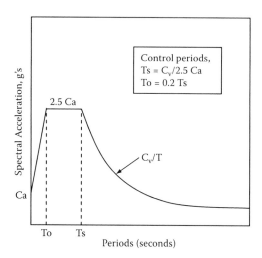

FIGURE 3.10   Design response spectra.

- External beam and the stair beams have width not less than 250 mm.
- The ratio between the building height to minimum horizontal direction is not more than 2.5.

There are different methods to calculate the earthquake loads on the structures and these methods are as follows:

1. Method of equivalent static applies to structures with uniform shape and structure systems whose vertical elements are continuous down to the foundation and does not exhibit sudden changes in stiffness. In addition, the building height is not more than 100 m and the maximum ratio of height to the horizontal dimension is equal to or more than 5 in the direction of the earthquake force.

$$V = Z \cdot I \cdot K \cdot C \cdot S \cdot W \tag{3.33}$$

where

Z is the earthquake intensity factor and its value is 0.1 in zone 1, 0.2 in zone 2, and 0.3 in zone 3.

I is the important factor and its value is as follows:

$I = 1.25$ in rescue, emergency, police stations, and fire fighting buildings

$I = 1$ for residential, public, and administration buildings

The K factor is obtained from Table 3.24.

The C factor is calculated from the following equation:

$$C = \frac{1}{15}\sqrt{T} \tag{3.34}$$

This factor has a value not more than 0.12.

TABLE 3.24

K Factors for Structure Systems

Structure System	K
All lateral loads carried by the shear walls or frame	1.33
Frame system to resist moment due to lateral load:	
• Ductile frame	0.67
• Nonductile frame	0.80
Double system consists of frame and shear wall and design based on:	1.00
• Frame and shear walls sharing in resisting lateral load according to its rigidity	
• Shear wall or frame separately resist lateral load	
• Frame system alone resists 25% of lateral load	

T is the fundamental natural frequency for the building in seconds, and can be obtained by performing a test on a similar structure or calculated by a root method. As an alternative one can define T for a high-rise building as follows:

a. For a structure with frame resisting moment and design to carry the total lateral load:

$$T = 0.1N \tag{3.35}$$

N is the number of floors including basement.

b. For multi-story buildings with different structure systems:

$$T = 0.09H \Big/ \sqrt{B} \tag{3.36}$$

H is the total building height and B is the horizontal direction of the building in the direction of earthquake effect. S is the soil factor and depends on the type of soil under the shallow foundation or under the pile tip (Table 3.25). W is the design weight of the structure and is taken equal to sustain dead load in a building that has a live load less than 500 kg/m^2 or equal to the dead load, and add half of the live load in the case of live load more than 500 kg/m^2.

TABLE 3.25

Soil Coefficient, S

S	Soil Type
1	Rock, dense soil, or very coherent with depth over 15 meters or soil medium density or coherent with depth of less than 15 meters above layer with best properties
1.15	Soil medium-density or coherent with depth deeper than 15 meters or loose soil or weak with depth less than 15 meters above layer with best properties
1.2	Loose soil or weak coherence of deeper than 15 meters

Distribute the total lateral force (V), which is calculated from the previous equation, to lateral static forces acting at the level of each slab in the building including the roof slab, and these forces calculate from the following equation:

$$F_j = \frac{W_j \cdot H_j \cdot (V - F_t)}{\sum\limits_{i=1,N} (W_i \cdot H_i)} \tag{3.37}$$

where
W_i = the design weight of the floor (i).
H_j = the height of the slab of floor (J) measured from foundation level.
F_t = the additional force affecting the roof slab level and calculated from the following equation:

$$F_t = 0.07T \cdot V \tag{3.38}$$

1. The value of F_t will be not more than 25% of the value of (V) and take equal to zero for T is less than or equal to 0.7 seconds.
2. The spectrum analysis method is for the building with uniform shape and structure system and building height between 100 and 150 m or the ratio between the heights to the horizontal dimension is more than 5 in the earthquake load direction. The effect of seismicity on the structure in this item is a static lateral load effect on the floor level of the building including the roof slab level and its value is calculated from the dynamic properties as the natural period and natural mode calculated by the modal analysis. The calculated lateral force should not be less than 80% of the lateral load calculates from the previous method.
3. Dynamic analysis applies to buildings with nonuniform shape or structure systems or height above 150 m or buildings that have changes of floor area to the next 25% or buildings that have eccentricity in design more than 25% of the higher horizontal dimension perpendicular to the lateral force. This method calculates the internal forces on the different internal elements to define the dynamic response with the earth movement due to earthquake by integrating the building motion with time. The dynamic analysis should include the dynamic properties for the structure including foundation and the soil carrying it.

Limits for lateral movement and joints:

- The horizontal movement between any two adjoining floors due to earthquake is not more than 5/1000 of the difference between levels of two floors.
- The space between two adjacent buildings or the width of separation between two parts in the same buildings more than 20 mm or the horizontal movement due to earthquake load.

3.3 COMPARISON OF CALCULATIONS OF RESISTANCE FOR DIFFERENT CODES

Every code uses different factors to reduce the strength, which is calculated from the design equations to cover the strength variation based on different quality control, materials properties, contractor's qualifications, and also the type of supervision.

In ACI, a factor less than unity called under-capacity reduction is used, where a partial safety factor is used in ECP and the British code.

Different partial safety factors are used to reduce the resistance of steel reinforcement and the concrete strength. ECP is in compliance with BS8110 in using the same partial factor values.

3.3.1 CAPACITY REDUCTION FACTORS

The reliable or usable provided strengths are obtained by multiplying the theoretical or nominal strengths by factors less than unity called under-capacity or capacity reduction factors. Since these factors are applied to the nominal strength, which represents the provided strength, a decrease in the value of a capacity reduction factor is equivalent to an increase in the overall factor of safety.

Capacity reduction factors are intended to allow for sources of uncertainty such as variations in materials properties, concrete dimensions, area of reinforcement bars and their location within the members, and other constructional inaccuracies.

According to clause 9.3.2 of the ACI code, the value of the capacity reduction factors in the case of axial compression tied members is equal to 0.7. According to Egyptian code ECP, the capacity reduction factors are presented by partial safety factors for concrete and steel as the British standard BS8110.

We can calculate the capacity reduction factor as follows in the case of columns:

$$Po = Pc + Ps \tag{3.39}$$

where

$$Pc = 0.67 \ Fcu \ Ac$$
$$Ps = As \ Fy \tag{3.40}$$

Equation (3.39) may be rewritten as follows:

$$Pc = \xi \, P_O \qquad Ps = (1 - \xi) \, P_O$$

where

$$\xi = \frac{1}{\left(1 + \rho \dfrac{Fy}{(0.67Fcu)} \right)} \tag{3.41}$$

where ξ is the percentage of the load carried by concrete to the total applied load to the column.

Introducing the strength reduction factors given by the ECP, the design equation for a short column is as follows:

$$P_0' = A_c \frac{0.67 F_{cu}}{\gamma_c} + A_s \frac{F_y}{\gamma_s} \tag{3.42}$$

where

γ_c = concrete strength partial reduction factor = 1.75.
γ_s = steel strength partial reduction factor = 1.36.

Substituting for γ_c = 1.75 and γ_s = 1.36 in equation (3.42) we have

$$P_0' = 0.38 A_c F_{cu} + 0.74 A_s F_y \tag{3.43}$$

EC code reduces the axial column strength by 10% for small eccentricity of columns (< 0.05t); therefore, equation (3.43) becomes

$$P_0' = 0.35 A_c \cdot F_{cu} + 0.67 A_s \cdot F_y \tag{3.44}$$

The strength equation (3.44) for columns can be written as

$$P_0' = 0.52(0.67 F_{cu} \cdot A_c) + 0.67(\rho \cdot A_c \cdot F_{cu})(F_y / F_{cu}) \tag{3.45}$$

Introducing an equivalent strength reduction safety factor, the axial load capacity of the column may be written as follows:

$$P_0' = \phi P_0 \tag{3.46}$$

Equating equations (3.45) and (3.46), the equivalent axial strength reduction factor for the column can be obtained as follows:

$$\phi = 0.52 \xi + 0.67 (1 - \xi) \tag{3.47}$$

So in equations (3.47) and (3.41) the capacity reduction factor in Egyptian code is variable and a function of percentage of longitudinal steel bars and the ratio between steel yield strength to concrete compressive strength.

The equivalent reduction factor in Egyptian code is calculated for different percentage of longitudinal steel bars and at various concrete grades as shown in Figure 3.11.

The maximum value of the equivalent reduction factor in Egyptian code is about 0.6, which is less than that of the ACI code. Therefore, the Egyptian code is more conservative in the case of resistance safety factor.

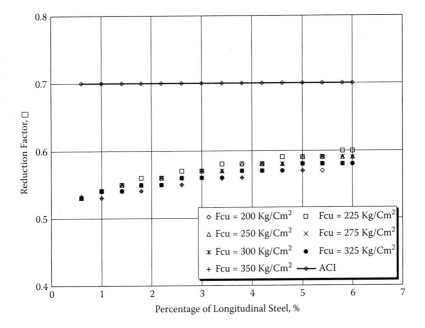

FIGURE 3.11 Comparison of reduction factor in EC and ACI for different percentages of steel and concrete grade.

3.3.2 PROBABILITY OF STRUCTURE FAILURE

Loads when calculated in specifications are taken as one value, but in fact the calculation is a probabilistic distribution. Calculating the member resistance capacity from code equation provides one value. Hence our calculation is called a deterministic calculation, but in the case of taking every parameter as variable presented by a probability distribution, the member resistance capacity will be a probabilistic distribution and the calculation is called a probabilistic calculation.

From Figure 3.12 one can see that the probability of structure failure occurs in the zone of intersection between the two curves of load and resistance. The failure

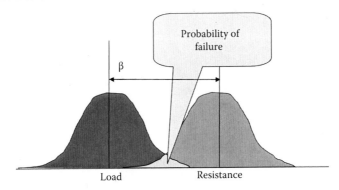

FIGURE 3.12 Structure reliability scheme.

of structure may be a partial collapse, total collapse, or member failure. This failure happens when the load value is at its highest and the strength is at its lowest value due to deterioration, poor quality control, or incorrect design.

From a practical point of view, the load increases due to a change in mode of operation or there are loads collecting at same time as, for example, collecting all furniture in one room during painting, or the load increases during parties.

The strength is weak usually due to bad quality control during construction or concrete deterioration due to corrosion in the steel bars and spalling concrete cover or cracks in concrete structure member in old buildings.

3.3.3 Probability-Based Limit States Design

The general limit states design philosophy (Allen 1976) has been adopted in the development of probability based design criteria. A limit state is a condition where a structure or structural element becomes unfit for its intended purposes. For most structures, limit states can be grouped into two categories:

1. Ultimate limit states, which are related to the structural failure or collapse of part or all of the structure, may lead to loss of life or major financial loss, and should have a low probability of occurrence.
2. Serviceability limits states, which are related to disruption of the functional use of the structure or structure damage or deterioration.

Limit states design is a process that involves: (a) the identification of all ways in which the structure might fail to fulfill its intended purpose; (b) the determination of acceptable levels of safety against each limit state; and (c) considerations by the designer of each significant limit state.

3.3.4 Reliability Analysis Procedure

Probability-based limit states design customarily employs loads or load effects which are multiplied by load factors and resistance which are multiplied by resistance factors in a set of checking equations which have the general form

$$\text{Factored resistance} \geq \text{effects of factored load}$$

The conceptual framework for probability-based design through this study is provided by classical reliability theory (Ang and Cornell 1974).

The loads and resistance are assumed to be statistical variables and the necessary statistical information is assumed to be variable. While this approach furnishes a sound theoretical basis for the evaluation of structural reliability, a number of conceptual and operational difficulties in its use have led to the development of first-order, second-moment (FOSM) methods, so called because of the way they characterized the uncertainty in the variables and the linearization underlying the reliability analysis (Galamos et al. 1982).

A mathematical model is first defined, which relates resistance and load effect variables for the specific limit state of interest:

$$g(X_1, X_2, \ldots\ldots X_m) = 0 \tag{3.48}$$

in which X_i = resistance or load variables, and failure is said to occur when $g < 0$ for the specific ultimate or serviceability limit state. Then the reliability associated with any given design is checked by identifying a point $(X_1, X_2, \ldots X_m)$ satisfying equation (3.48) such that (Galambos et al. 1982)

$$x_i = \mu_{xi} - \alpha_i \beta \sigma_{xi} \tag{3.49}$$

$$\alpha_i = \frac{\sigma_{xi} \dfrac{\partial g}{\partial x_i}}{\sqrt{\sum\left(\sigma_{xi} \dfrac{\partial g}{\partial x_i}\right)^2}} \tag{3.50}$$

where, μ_{xi} and σ_{xi} are the mean and standard deviation, respectively, and β is defined as the reliability index. The partial derivatives in equation (3.50) are evaluated at the point defined by equation (3.49). This point is called checking or design point. It is found that (Ang and Tang 1984) searching iteratively for the direction cosines of this point, αi, that minimize β, where β is the shortest distance from the origin to the failure surface.

According to Ang and Tang (1984), the linear performance function may be represented as

$$g(X) = a_o + \sum_i a_i X_i \tag{3.51}$$

where a_o and a_i's are constants. The shortest distance from the origin to the failure plane surface will be

$$\beta = \frac{a_o + \sum_i a_i \mu_{xi}}{\sqrt{\sum_i (a_i \sigma_{xi})^2}} \tag{3.52}$$

When information on the probability distributions of the design variables in equation (3.48) is available, in addition to estimates of their statistical parameters (mean and variances), this information may be included in the analysis in an approximate way that does not entail the multidimensional integrations required in classical reliability analysis.

Many structural problems involve random variables that are clearly nonnormal (Corotis et al. 1980, Simiu 1975). Nonnormal variables are transformed to equivalent normal variables to the solution of equations (3.49), (3.50), and (3.51).

Transformation should take place at a point close to that where failure is most likely, that is to say, at a point on the failure surface; herein μ_{Xi} and σ_{xi} of the equivalent normal variable are chosen so that at the checking point, Xi, the cumulative probability distribution and probability density function of the actual and approximately normal variable are equal (Ang and Tang 1984). Having determined μ_{Xi} and σ_{xi} of the equivalent normal distributions, the solution proceeds as in equations (3.50) and (3.51).

3.3.5 CALCULATION OF RELIABILITY INDEX

Applying the previous procedure for calculating reliability on the ECP limit state equation in the case of reinforced concrete column, the equation of limit state is as follows:

$$\phi \, Rn \geq 1.4 \, Dn + 1.6 \, Ln \qquad (3.53)$$

where Rn, Dn, and Ln are respective nominal values of resistance and loads.

Resistance reduction factor (ϕ) will be calculated, as previously discussed in section 3.3.1, depending on the percentage of longitudinal steel bars and the ratio between yield strength and concrete compressive strength.

Structural engineers often use "nominal" values of loads and resistances, which are generally different from corresponding mean value. For example, according to Galambos et al. (1982), the ratios of the mean loads to the respective specific nominal loads are as follows:

$$\frac{\mu_D}{Dn} = 1.05$$

$$\frac{\mu_L}{Ln} = 1.15$$

whereas in the case of axial compression of tied column:

$$\frac{\mu_R}{Rn} = 0.95$$

Assume R & D are normally distributed random variables and L is a type I asymptotic random variable. The corresponding coefficients of variation associated with these variables (Galambos et al. 1985) are as follows:

$$VR = 0.11$$

$$VD = 0.10$$

$$VL = 0.25$$

Assume Ln/Dn = 1.0 which is a live load to dead load ratio.

$$\frac{\mu_L / 1.15}{\mu_D / 1.05} = 1.0 \qquad \mu_L = 1.095 \, \mu_D$$

whereas the ECP requirement becomes

$$\varphi \frac{R^`}{0.95} = 1.4 \frac{D^`}{1.05} + 1.6 \frac{L^`}{1.15}$$

Assume $\varphi = 0.54$, which corresponds to $\rho = 1\%$ as discussed in section 3.3.1.

The cumulative and density probability asymptotic type I random variables will be as follows:

$$F_L(l) = \exp(-e^{-\alpha_L(l-u_L)})$$

$$f_L(l) = \alpha_L \exp(-\alpha_L(l-u_L) - e^{-\alpha_L(l-u_L)})$$

The parameters that describe the asymptotic live load α_L and u_L will be calculated as follows:

$$\alpha = \frac{\pi}{\sqrt{6}} \cdot \frac{1}{\sigma_L} = \frac{4.686}{D^`}$$

$$u = L^` - (0.577/\alpha) = 0.9721 \, D^`$$

In this case, that is, involving nonnormal distributions, even though the performance function is linear, the mean values and standard deviations are unknown as these are now functions of respective failure point values. An iterative solution using equations (3.17) and (3.18) is therefore necessary.

For the first iteration assume

$$l^* = \mu_L = 1.095 \, \mu_D$$

For the type I extremal distribution of L

$$F_L(l^*) = 0.57$$

$$f_L(l^*) = 1.5/D^`$$

and thus from equations (3.17) and (3.18) to transform from nonnormal random variables to equivalent normal variables

$$\sigma_L^N = 0.26\mu_D$$

$$\mu_L^N = 1.05\mu_D$$

Knowing the value of the mean and standard derivation of the normally equivalent normally distributed variables, the safety index can be calculated using equation (3.52) and it is found to be:

$$\beta = \frac{\mu_R^N - \mu_D^N + \mu_L^N}{\sqrt{(\sigma_R^N)^2 + \sigma_D^2 + (\sigma_L^N)^2}} = 4.18 \qquad (3.54)$$

and the failure point is

$$D^* = \mu_D - \beta\sigma_D$$

$$R^* = \mu_R - \beta\sigma_R \qquad (3.55)$$

$$L^* = \mu_L^N - \alpha_L^*\beta\sigma_L^N$$

where

$$\alpha_L^* = \frac{-\sigma_L^N}{\sqrt{(\sigma_R^N)^2 + \sigma_D^2 + (\sigma_L^N)^2}}$$

The value of L^* calculated from equation (3.16) is used to another iteration to improve the value of the safety index β.

After some iteration one can get the reliability index $\beta = 4.1$ which corresponds to the ECP requirements for columns using $Ln/Dn = 1.0$ and $\rho = 1\%$.

Therefore, from the above procedure one can calculate the reliability index of the Egyptian code ECP in the case of reinforced concrete tied column.

From Figure 3.11, one can obtain the reliability index of the Egyptian code ECP in the design of reinforced concrete column, which corresponds to different percentage of longitudinal steel bars.

Moreover, it is noticed that the reliability index has a higher value than that for the ACI code, in which the reliability index has a constant value of 3.4 (Galambos 1982) for all percentages of steel bars.

One can observe from Figure 3.11 that the reduction factor increases as the percentage of steel bars increases, while the reliability index decreases as the percentage of steel increases.

In the case of a different live load to dead load ratio, the reliability index is calculated at different percentages of steel bars (0.6%, 3%, and 6%) and plotted as shown in Figure 3.13.

From this figure, the relation between the live load to dead load ratio to the reliability index for different percentage of steel is shown and one can obtain that at higher live load to dead load ratio; the reliability index is higher for Egyptian code.

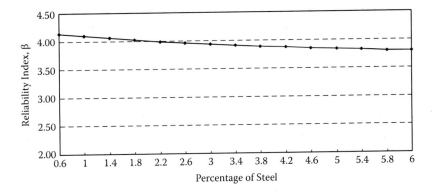

FIGURE 3.13 Comparison of EC code and ACI code reliability indices for different percentages of longitudinal steel.

3.3.6 EFFECT OF C.O.V. OF RESISTANCE ON SAFETY INDEX

The variation on the resistance is dependent on the variation of material properties, concrete dimensions and steel bars areas, and the variation of uncertainty in the theoretical equation of column design (MacGregor et al. 1983).

Therefore, in the following analysis it is required to calculate the mean resistance strength and coefficient of variation (C.O.V.) by knowing the mean and C.O.V. levels of the different variables of the column cross section.

As discussed in the analysis of the capacity reduction factor in section 3.3.1 the resistance of column is carried by concrete and steel strength.

$$Po = Pc + Ps$$

$$Po = \lambda c\, Pc + \lambda s\, Ps$$

where λ is the ratio between the mean value and the nominal value for the parameter.

From analysis at section 3.3.1 one can get

$$Po = \lambda c\, \xi\, Po + \lambda s\, \xi\, Po \tag{3.56}$$

where

$$\lambda c = \frac{\mu_{Fcu}}{Fcu} \cdot \frac{\mu_{Ac}}{Ac} \tag{3.57}$$

$$Ac = A \times B$$

$$\lambda s = \frac{\mu_{Fy}}{Fy} \cdot \frac{\mu_{As}}{As} \tag{3.58}$$

where A, B are the dimensions of the reinforced concrete column and ξ, which is the percentage of the load carried by concrete to the total applied load to the column, is calculated from equation (3.41).

$$\mu_R = \mu_{Po} \times \mu_P \tag{3.59}$$

where μ_P is the accuracy of design equation due to use of the rectangular stress block, the limiting strain, and the neglect of strain hardening. For tied columns, Mattock (1961) found mean ratios of test to calculate strengths ranging from 0.97 to 1.00 with the C.O.V. ranging from 0.046 to 0.074 including possible in-test variations. In this study the mean ratio of actual strength to design strength will be taken as $P = 0.98$ and C.O.V. as $V_P = 0.05$.

$$\gamma_{\bar{R}} = \frac{\mu_{Po}}{Po} = (\lambda c \cdot \xi + (1 - \xi) \cdot \lambda s) \cdot \mu_P$$

$$\varphi = (\gamma_{Rc} + \gamma_{Rs}) \cdot \mu_P \cdot e^{-\beta \alpha V_R} \tag{3.60}$$

$$V_{Po}^2 = V_R^2 - V_P^2$$

$$\sigma_{Pc} = V_{Pc} \cdot \lambda c \cdot Pc \tag{3.61}$$

$$\sigma_{Ps} = V_{Ps} \cdot \lambda s \cdot Ps$$

$$\sigma_{Po} = \left[\sqrt{(V_{Pc}\lambda c \xi)^2 + (V_{Ps}\lambda s(1 - \xi))^2} \right] Po \tag{3.62}$$

From equations (3.61) and (3.62) one can obtain the C.O.V. as follows:

$$V_{Po} = \frac{\sqrt{(V_{Pc}\lambda c \xi)^2 + (V_{Ps}\lambda s(1 - \xi))^2}}{\lambda c \cdot \xi + \lambda s \cdot (1 - \xi)} \tag{3.63}$$

$$V_R = \sqrt{V_{Po}^2 + V_P^2} \tag{3.64}$$

$$V_R^2 = \frac{(V_{Pc}\lambda c \xi)^2 + (V_{Ps}\lambda s(1 - \xi))^2}{(\lambda c \cdot \xi + \lambda s \cdot (1 - \xi))^2} + V_P^2 \tag{3.65}$$

where

$$V_{Pc}^2 = V_c^2 + 2V_A^2$$

$$V_{Ps}^2 = V_{Fy}^2 + V_{As}^2$$

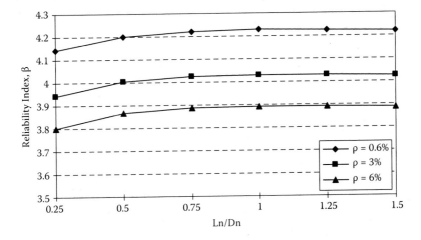

FIGURE 3.14 Comparison of the reliability index and live load to dead load ratio for different percentages of steel.

To study the effect of resistance variation on the safety index of ECP code, different values of C.O.V. of resistance are assumed (0.12, 0.14, 0.16, and 0.20).

The reliability index is calculated corresponding to the assumed values of strength C.O.V. and at different live load to dead load ratio and plotted in Figure 3.14.

From this figure one can conclude that at higher C.O.V. of resistance strength of reinforced concrete column the ECP code has a lower reliability index than that in the case of lower resistance C.O.V. (Figure 3.15).

From this study, for any values of resistance capacity C.O.V. when increasing the ratio of live load to dead load, the ECP code provides an increase in the reliability index.

CV summary, by discussing the comparison between live load in ECP code to that for different codes, one can see that ECP is more conservative.

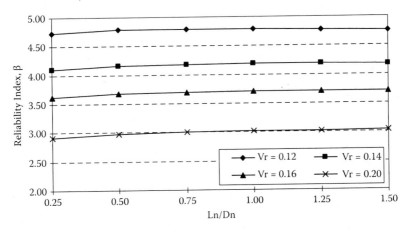

FIGURE 3.15 Live load to dead load ratio versus reliability index in EC for different strength coefficients of variation.

On the other hand, in discussing the strength capacity reduction factor for ECP and ACI, it is observed that the Egyptian code is more conservative.

By comparing the reliability index for Egyptian code to that for ACI, it is found that the Egyptian code provides a higher safety index than the other.

Moreover, by knowing the variation of different variables effects on resistance strength, one can calculate the mean and C.O.V. of the resistance strength.

REFERENCES

Allan, D. E. 1976. Limit States Design: A Probabilistic Study. *Canadian Journal of Civil Engineering.* 2(1):36–49.

American National Standards Institute. 1982. *Building code requirements for minimum design loads in buildings and other structures, ANSI A58.1-1982.* New York, NY.

Ang, A. H., and W. H. Tang. 1984. *Probability concepts in engineering planning and design. Vol. II. Decision, risk and reliability.* New York, NY: John Wiley & Sons, Inc.

Ang, A. H., and Cornell, C. A. Reliability Based Structural Safety and Design, *J. of Struct. Div.,* ASCE, Vol.100, No. ST9, Sept. 1974. Proc. Paper 10777, 1755–1769.

Borges, J., and Castanheta, M., 1972 Statistical Definition Of Combination Of Loads, probabilistic design of reinforced concrete building, publication, SP-31, American Concrete Institute, Detroit, MI.

Bosshard, W. 1975. *On stochastic load combination. Technical report no. 20.* Stanford, CA: J. A. Blume, Earthquake Engineering Center.

Cornel, C. A. and Benjamin, J.R., 1974, *Probability, statistics and decision for civil engineers,* McGraw-Hill, Inc., New York.

Corotis, R. B., and P. L. Chalk. 1980. Probability models for design live loads. *Journal of the Structural Division.* ASCE. 106(ST10):2017–2033.

Corotis, R. B., and V. A. Doshi. 1980. Probability models for live load survey results. *Journal of the Structural Division.* ASCE. Proc. paper 18783. 106(ST10):2017–33.

Corotis, R. B., and W-Y. Tsy. 1983. Probabilistic load duration model for live load. *Journal of the Structural Division.* ASCE. 109(4):859–73.

Crystal Ball Program version 3.0 user manual. 1996. Decision Engineering.

Der Kiureghiam, A. 1978. Second-Moment Combination of Stochastic Loads. *Journal of the Structural Division.* ASCE. Proc. paper 14056,(ST10):1551–1567.

Der Kiureghiam, A. 1980. Reliability analysis under stochastic load. *Journal of the Structural Division.* ASCE. Proc. paper 18190. 106(ST2):411–29.

El-Reedy M. A., M.A. Ahmed, and A.B. Khalil. (2000). Reliability Analysis of Reinforced Concrete Columns. Ph.D. thesis, Cairo University, Giza, Egypt.

Galamobos, T. V, Ellingwood, B, MacGregor, and Cornell, C. A, Probability Based Load Criteria: Assessment of Current Design Practice, ASCE, Vol.108, No. ST5, May 1982, Proc.17067, pp. 959–977.

Grigoriu, M. 1975. *Process load on the maximum of sum of random models. International document no. 1.* Structural Loads Analysis and Specification Project. Department of Civil Engineering. Cambridge, MA: Massachusetts Institute of Technology.

Harris, M. E., R. B. Corotis, and C. J. Bova. 1981. Area dependent process for structural live load. *Journal of the Structural Division.* ASCE. 107(STS):857–71.

Larrabee, R. 1978. *Approximate stochastic analysis of combined loading.* Structural Science Analysis and Specification Project. Department of Civil Engineering. Cambridge, MA: Massachusetts Institute of Technology.

MacGregor, J. G., Load and Resistance Factors for Concrete Design, ACI J., Vol. 80(4), 1983 pp. 279–287.

Mattock, A.H., Kriz, L. B. and Hongnestad, E.,Rectangular Concrete Stress Distribution in Ultimate Strength Design, Proc., ACI, Vol.57, Feb. 1961, pp. 875–928.

Meguire, R. K., and C. A. Cornell. 1974. Live load effects in office buildings. *Journal of the Structural Division*. ASCE. Proc. paper 10660. 100(ST7).

MicroSoft Excel spread sheet Rev. 2006.

National building code. 1976. New York, NY: America Assurance Association.

Parzen, E. 1967. *Stochastic processes*. San Francisco, CA: Holden Day.

Person, R. 1993. *Using EXCEL version 5 for Windows*. Que Corporation.

Pier, J. C., and C. A. Cornell. 1973. Spatial and temporal variability of live load. *Journal of the Structural Division*. ASCE. Proc. paper 9747. 99(STS):903–22.

Simiu, E., and Filliben, J. J., Practical Approach to Code Calibration, J. Struct. Div., ASCE, Vol.110(ST7), 1975, pp. 1469–1480.

Wolff, R. W. 1989. *Stochastic modeling and the theory of queues*. New York, NY: Prentice-Hall International.

Wen, Y. K. 1977. Statistical combination of extreme loads. *Journal of the Structural Division*. ASCE. Proc. paper 12930. 103(STS):1079–93.

4 Concrete Materials and Tests

4.1 INTRODUCTION

Concrete consists of two main parts. The first part is the aggregate (coarse and fine) and the other part is the adhesive between cement and water. The main components are the coarse and fine aggregate, cement, water, and additives that improve the properties of concrete. These additives must be within some definite percentage as they create a negative impact if the dosage is not accurate.

To obtain a concrete that conforms to the specifications, we must adjust the quality of each component in the concrete mix. Therefore, in this chapter the main necessary tests to cement, aggregate, water, and additives will be illustrated to obtain concrete with a high quality and to match the international standard and project specifications.

Concrete is weak under tensile stress, so we use steel reinforcement to withstand the tensile stress, which increases the efficiency of concrete. Nowadays reinforced concrete is an important element in the construction industry worldwide because it is cheaper in most cases than other alternatives and because of its ease of formation in the early stages, which allows different architectural forms.

Steel is the most important element in reinforced concrete structures as it carries most stresses. Steel bars must be under a strict system of quality control to make sure they meet international standards and project specifications.

These tests should be done with valid calibration devices to give accurate results. The laboratory should follow a quality assurance system and the devices should be calibrated periodically. Moreover, the samples must be taken correctly according to standard specifications and all the necessary tests performed and review based on the quality system.

4.2 CONCRETE MATERIALS TEST

The following sections describe the important field and laboratory tests for concrete materials as most civil engineers on site focus on the compressive strength and slump test only and do not pay attention to other tests.

Note that the compressive strength and slump test is important for concrete durability and predicting performance with the time. Moreover, for cube or cylinder compressive strength, it gives results 28 days after concrete has become hard so the modification or rebuild is very difficult. Materials tests ensure that the concrete strength will achieve the project specification target.

4.2.1 CEMENT

The use of materials for adhesion is very old. The ancient Egyptians used calcined impure gypsum. The Greeks and the Romans used calcined limestone and later learned to add sand and crushed stone or brick and broken tiles to lime and water. Lime mortar does not harden under water, and for construction under water the Romans ground together lime and volcanic ash or finely ground burnt clay tiles. The active silica and alumina in the ash and the tiles combined with lime to produce what became known as pozzolanic cement from the village of Pozzuoli, near Mount Vesuvius, where the volcanic ash was first found. The name of pozzolanic cement is used to this day to describe cements obtained simply by the grinding of natural material at normal temperature.

Nowadays, there are different types of cement and choosing alternatives depends on the designer's recommendations in the project specifications, the drawings based on the surrounding enviromental conditions, and the required strength.

The main types of cement in British standards and those of the American Society for Testing and Materials (ASTM) are shown in Table 4.1.

To ensure cement quality, there are many tests to define whether the quality of the cement matches the specifications. There are some important tests that are applied in many international standards and define the limits of cement as stated in British standards and ASTM.

There are many ways to take a sample according to the nature of the cement required by the procurement agreement that may call for cement loose or in bags on site.

It is very important to define in the agreement between the supplier and the customer the location and time to take the samples, and allow enough time to take the samples to perform the required tests, obtain the results, and define the suitability of the cement.

Testing should be performed within 28 days of receiving the samples.

TABLE 4.1
Cement Types

ASTM Description	British Description
Type I	Ordinary Portland cement
Type III	Rapid hardening Portland
	Extra rapid hardening Portland
	Ultra high early strength Portland
Type IV	Low heat Portland
Type II	Modified cement
Type V	Sulfate resisting Portland
Type IS	Portland blast furnace
	White Portland
Type IP & P	Portland pozzolana
Type S	Slag cement

When taking the samples, keep in mind that they must conform with the specifications agreed upon. The general requirement in the standard specification is that the weight of one sample is not less than 5 kg and that the sampling tools are dry and clean. The sample may be taken individually via digital sampling of bulk cement and there may be a number of samples spaced at intervals, which is called composite sampling. In the case of bulk cement try to avoid the upper cement layer by about 150 mm from the top.

Most sites receive the cement in sacks so the tests will done by selecting random sacks and the number of sacks tested will be according to the following equation:

$$\text{The number of samples} \geq (n)^{0.333}$$

4.2.1.1 Cement Test by Sieve No. 170

The fineness of cement affects the quality of the concrete industry in general. A big cement particle cannot completely react with water as water cannot reach a remaining core in the cement particle. The water propagates through the cement particles and they start to dehydrate, which causes an increase in temperature, which is the main reason for the forming of hair cracks and preventing stabilization of cement volume.

As a result, an increase in the cement particle size reduces the strength of the same cement content and increasing the fineness of the cement will improve the workability, cohesion, and durability with time and decrease the water moving upward to the concrete surface.

Figure 4.1, from Neville's book (1983), presents the relation between concrete strength and the concrete fineness at different ages. To perform this test, take a sample of 50 g of cement and shake it in a closed glass bottle for two minutes and then revolve the sample gently using dry bar. Put the sample in a closed bottle and leave it for two minutes. Put the sample in 170 sieve (90 microns) and move it, shaking the sieve horizontally and rotationally, then confirm finishing the sieve test when the rate of passing cement particles is not more than 0.5 g/min during the sieve process. Remove the fines carefully from the bottom of the sieve using a smooth brush. Then, collect and weigh the remaining particles on the sieve (W1).

Repeat the same test with another sample. Then the residual weight for the second test is obtained (W2). Calculate the values of the remaining samples through

$$R1 = (W1/50) \times 100$$

$$R2 = (W2/50) \times 100$$

The ratio (R) is calculated by taking the average of R1 and R2 to the nearest 0.1% and, in the case of deviating results of the two samples, more than 1%. Do the test a third time and take the average of the three results.

You can accept or refuse the cement based on the following condition:

- For Portland cement the R must not exceed than 10%.
- For rapid hardening Portland cement the R must not exceed 5%.

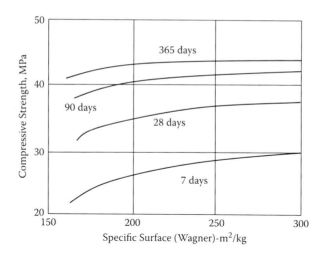

FIGURE 4.1   Cement fines and compressive strength.

4.2.1.2   Initial and Final Setting Times of Cement Paste Using Vicat Apparatus

The objective of this test is to define the time for initial and final setting of the paste of water and cement with standard consistency by using a Vicat apparatus and determine whether the cement is expired or can be used.

The initial setting is the required time to set and after that concrete cannot be poured or formed; the final setting time is the time required for the concrete to be hardened.

Vicat apparatus (Figure 4.2) consists of a carrier with needle acting under a prescribed weight. The parts move vertically without friction and are not subject to erosion or corrosion. The paste mold is made from a metal or hard rubber or plastic like a cut cone with depth of 40 ± 2 mm and the internal diameter of the upper face 70 ± 5 mm and lower face 80 ± 5 mm and provides a template of glass or similar materials in the softer surface. Its dimensions are greater than the dimensions of the mold.

The needle is used to determine the initial setting time in a steel cylinder with effective length 50 ± 1 mm and diameter 1.13 ± 0.5 mm. The needle measuring time is in the form of a cylinder with length of 30 ± 1 mm and diameter 1.13 ± 0.5 mm and held by a 5 mm diameter ring at the free end to achieve distance between the end of the needle and the ring of 0.5 mm.

The test starts by taking a sample weighing about 400 g and placing it on an impermeable surface and then adding 100 ml of water and recording zero measurement from the time of adding water to the cement and then mixing for 240 + 5 seconds on the impermeable surface.

To determine the initial setting time and calibrate the device until the needle reaches the base of the mold, then adjust the measuring device to zero and return needle to its original place.

Fill the mold with cement paste with standard consistency and troll the surface, then put the mold for a short time in a place that has the the temperature and humidity required for the test.

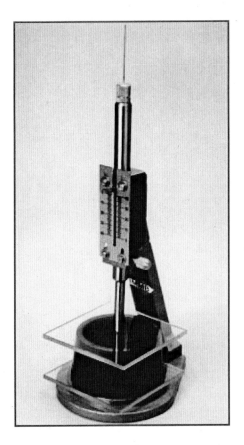

FIGURE 4.2 Vicat apparatus.

Transfer the mold to the apparatus under the needle, and then make the needle slowly approach the surface until it touches the paste's surface, stop it in place for a second or two seconds to avoid impact of primary speed, then allow the moving parts to implement the needle vertically in the paste.

Grading depends on when the needle stops penetrating or after 30 seconds, whichever is earlier, and indicates the distance between the mold base and the end of the needle, as well as the time start from the zero level measurement.

Repeat the process of immersing the needle in the same paste in different locations with the distance between the immersing point and the edge of the mold or between two immersing points not less 10 mm after about 10 minutes, and clean the needle immediately after each test.

Record time is measured from zero up to 5 ± 1 mm from the base of the mold as the initial setting time to the nearest 5 minutes. Ensuring the accuracy of measurement of time between tests reduces embedment and the fluctuation of successive tests. The needle is used to identify the final time of setting; follow the same steps as in determining time of initial setting and increase the period between embedment tests to 30 minutes.

Record the time from zero measurement until embedment of the needle to a distance of 0.5 mm, which will be the final setting time. Control the impact of the needle on the surface of the sample so the final setting time presents the effect of the needle. To enhance the test's accuracy reduce the time between embedment tests and examine the fluctuation of these successive tests. Record the final setting time to the nearest 5 mm.

According to the Egyptian specifications the initial setting time must not be less than 45 minutes for all types of cement except the low heat cement, for which the initial setting time must not be less 60 minutes. The final setting time must be shorter than 10 hours for all types of cement.

4.2.1.3 Density of Cement

The purpose of this test is to determine the density of cement by identifying the weight and unit volume of the material by using the Le Chatelier density bottle. The determination of the cement density is essential for concrete mix design and to control its quality. This test follows specifications of the American Society for Testing and Materials, ASTM C188-84.

The Le Chatelier device is a standard round bottle. Its shape and dimensions are shown in Figure 4.3. This bottle must have all the required dimensions, lengths, and uniform degradation and accuracy.

The glass that is used in the Le Chatelier bottle must be of high quality and free from any defects. It should not interact with chemicals and have high resistance to heat and appropriate thickness to have a high resistance to crushing. Measurements start at the bottle's neck and go from zero to 1 mL and from 18 to 24 mL with accuracy to 0.1 mL. Each bottle must have a number to distinguish it from any other. Write on the bottle the standard temperature and the capacity in millimeters over the highest point of grading.

Processed sample cement weighing about 64 g to the nearest 0.05 g must be tested.

Fill the bottle with kerosene free from water and oil whose density is at least 62 API. Up to point gradations between zero and 1 mL, dry the inner surface of the bottle at the highest level of kerosene if necessary, and use rubber on the surface of the table used for the test when filling the bottle.

The bottle, which is filled with kerosene, is placed in a water bath and the first reading to kerosene level is recorded. To record the first reading correctly install the bottle in the water bath vertically. Put a cement sample weighing 64 g with accuracy to 0.05 g inside the bottle with small batches at the same temperature of kerosene, taking into account when putting the cement inside the bottle to avoid cement dropping out or its adhesion on the internal surfaces of the bottle at the highest level. The bottle can be placed on the vibrating machine when putting the cement inside the bottle to expedite the process and prevent adhesion of granulated cement with the internal surfaces of the bottle.

After laying the cement inside the bottle, put a cap on the bottle mouth and then spin diagonally on the surface so as to expel the air between the granules of cement, and continue moving the bottle until the emergence of air bubbles stops from the kerosene surface inside the bottle.

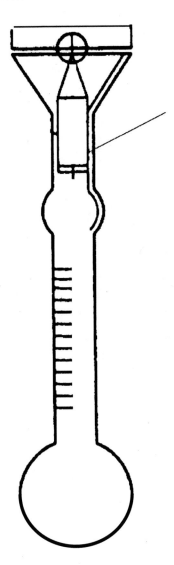

FIGURE 4.3 Le Chatelier bottle.

Put the bottle in the water bath and then take the final reading, and record the reading at the lower surface of kerosene so as to avoid the impact of surface tension. For the first and final readings, make sure that the bottle is placed in a water bath with constant temperature for a period not to exceed the difference in temperature between the first and final readings of about 0.2°C.

The difference between the first and final reading is the volume of the moving liquid by the cement sample.

The volume of the moving liquid = final reading – first reading

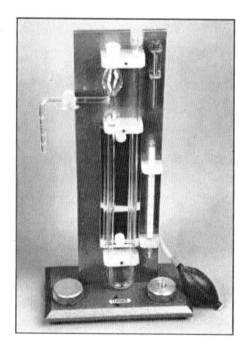

FIGURE 4.4   Blaine apparatus.

Cement density = cement weight (g)/moving liquid volume (cm³)

The accuracy of the density calculation is to the nearest 0.01 g/cm³.

4.2.1.4   Define Cement Fineness by Using Blaine Apparatus

This test is used to determine the surface area by comparing the test sample with the specific reference. The greater surface area increases the speed of concrete hardening and obtains early strength. This test determines the acceptance of the cement.

There are many tests to define cement fineness and one is a Blaine apparatus as stated in many codes such as the Egyptian code.

This test depends on calculating the surface area by comparing the sample test and the reference sample using a Blaine apparatus to determine the time required to pass a definite quantity of air inside a cement layer with defined dimensions and porosity.

A Blaine apparatus is shown in Figure 4.4. The first step in testing is to determine the volume of the cement layer using mercury in the ring device of the Blaine apparatus. Cement is then added and by knowing the weight of the cement before and after adding it as well as the mercury density, the volume of the cement layer can be calculated.

$$V = W_1 - W_2/Dm$$

where

V is the volume of cement layer, cm³.

W_1 is the weight of mercury in grams that fills the device to nearest (0.0 g).
W_2 is the weight of mercury in grams that fills the device to nearest (0.0 g).
Dm is the density of the mercury (g/cm³). From tables, define the mercury density at the average temperature of the test by using the manometer in the Blaine apparatus.

$$S_r = \frac{K}{D_r} \left(\frac{\sqrt{(P_r)^3 T_r}}{(1-P_r)\sqrt{0.1 I_r}} \right)$$

From the previous equation:

S_r is the reference cement surface area, (cm²/g).
D_r is the reference cement density (g/cm³).
P_r is the porosity of the cement layer.
I_r is the air visciosity in the average temperature for reference cement test.
T_r is the average time required for the manometer liquid to settle in two marks to nearest 0.2 sec.
K is the Blaine apparatus constant factor defined by the previous equation by knowing the time needed to pass the air in the sample.

To retest the sample, we calculate its surface area by using the following equation:

$$Sc = Sr(Dr/Dc) *(Tc/Tr)^{0.5}$$

According to the Egyptian code, the acceptance and refusal of cement is based on limites shown in Table 4.2.

TABLE 4.2
Cement Fineness Acceptance and Refusal Limits

Cement Types	Cement Fineness Not Less Than cm²/gm
Ordinary Portland	2750
Rapid hardening Portland	3500
Sulfate resistant Portland	2800
Low heat Portland	2800
White Portland	2700
Mixing sand Portland	3000
4100 fineness	4100
Slag Portland	2500

4.2.1.5 Compressive Strength of Cement Mortars

The cement mortar compressive strength test is performed using standard cubes of cement mortar mixed manually and compacted mechanically using a standard vibrating machine. This test is considered a refusal or acceptance determination.

Compressive strength is one of the most important properties of concrete. The concrete gains its compressive strength from cement paste as a result of the interaction between the cement and water added to the mix. So it is critical to make sure that the cement used is the appropriate compressive strength. This test should be done to all types of cement.

Needed for the test are stainless steel sieves with standard square holes opened 850 or 650 microns. Stainless steel does not react with cement and weighs 210 g.

The vibrating machine has a weight of about 29 kg and the speed of vibration is about 12,000 vertical vibrations + 400 RPM and the moment of vibrating column is 0.016 N.m.

The mold of the test is a cube 70.7 ± 1 mm, the surface area for each surface is 500 mm², the acceptable tolerance in leveling is about 0.03 mm, and the tolerance between paralleling for each face is about 0.06 mm.

The mold is manufactured from materials that will not react with the cement mortar, and the base of the mold is made from steel that can prevent leaks of the mortar or water from the mold. The base is matched with the vibrating machine.

The sand should contain a percentage of silica not less than 90% by weight and must be washed and dried very well. Moreover, the humidity of the sand must not be more than 0.1% by weight for it to pass through a sieve with openings of 850 microns, and for it to pass through the standard sieve size of 600 microns it should not have more than 10% humidity by weight (Tables 4.3 and 4.4).

After performing the tests, the standard cubes will be crushed within one day, which is about 24 ± 0.5 hours, and three days in the limits of 72 ± 1 hour, and after seven days within 168 ± 1 hour, and after 28 days within 672 ± 1 hour.

Table 4.5 illustrates the limits of acceptance and rejection according to the cement mortar compressive strength. Note from the table that there is more than one type of high-alumina cement as the types vary according to the percentage of oxide alumina. The compressive strength after 28 days will not be considered accepted or rejected unless clearly stated in the contract between the supplier and the client.

TABLE 4.3
One Cube Mixing Ratio

Cement Type	Ratios by Weight	Materials	Weight (g)
All types of cement	1.0	Cement	185 ± 1
	3.0	Sand	555 ± 1
	0.4	Water	74 ± 1
High alumina cement	1.0	Cement	190 ± 1
	3.0	Sand	570 ± 1
	0.4	Water	76 ± 1

TABLE 4.4
Allowable Deviation in Temperature and Humidity

Location	Temperature (°C)	Min. Relative Humidity (%)
Mixing room		65
Curing room		90
Water curing sink	20 ± 2	—
Compression machine room		50

TABLE 4.5
Acceptance and Refusal Limits

	Cube Compressive Strength (N/mm²)			
Cement type	After 28 days ≥	After 7 days ≥	After 3 days ≥	After 24 hours ≥
Ordinary Portland	—	18	27	36
Rapid hardening Portland	—	24	31	40
Sulfate resistant Portland	—	18	27	36
Low heat Portland	—	7	13	27
White Portland	—	18	27	36
Mixing sand Portland	—	12	20	27
4100 fineness	10	25	32.5	40
Slag Portland	—	13	21	34
80	25			
70	30			
50	50			
40	50			

4.2.2 Aggregate Tests

4.2.2.1 Sieve Analysis Test

The main key to obtaining higher concrete strength is the aggregates' interactions with each other. This interaction increases the density of concrete, which increases the compressive strength, and the interaction between aggregates depends on the grading of coarse aggregate and sand that should be done according to definite specifications. This test relies on standard sieves and knowledge of specifications and dimensions as in Table 4.6.

These standard sieves have metal cylinder frames with square openings. Each sieve is labeled by its opening length in millimeters and shape as shown in Figure 4.5.

The test procedure starts by defining the weight of the aggregate, and then drying it at a temperature of 105°C ± 5°C for 24 hours until the weight proves to the nearest 0.1%. Sieves are arranged according to size, and the sample is placed on the largest

TABLE 4.6
Standard Sieve Sizes for Aggregates Set by Various Standards, mm

EC	BS410:1986	BS812: 103.1:1985	EN 993-2	ASTM E11-87
	125			125
				100
	90			
75		75		75
63	63.0	63	63	
50		50		
	45			
37.5		37.5		37.5
	31.5			
		28		
26.5				
				25
	22.			
				4
		20		
19				19
	16		16	
		14		
13.2				
				12.5
	11.2			
		10		
9.5				9.5
	8.0		8.0	
6.7				
		6.3		6.3
	5.6			
		5.0		
4.75				4.75
	4.0		4.0	
3.35		3.35		
	2.8			
2.36		2.36		2.36
	2.0		2.0	
1.7		1.7		
	1.4			
1.18		1.18		1.18
	1.00		1.00	

TABLE 4.6 (*Continued*)
Standard Sieve Sizes for Aggregates Set by Various Standards, mm

EC	BS410:1986	BS812: 103.1:1985	EN 993-2	ASTM E11-87
0.85		0.85		
	0.710			
0.6		0.6		0.6
	0.5		0.5	
0.425				
	0.355			
0.3		0.30		0.30
	0.25		0.25	
0.212		0.212		
	0.18			
0.15		0.15		0.15
	0.125		0.125	
	0.090			
0.075		0.075		0.075
	0.063		0.063	
	0.045			
	0.032			

of the sieves. Start shaking the sieve manually or mechanically for at least 5 minutes to ensure that there does not pass from any sieve after that period more than 0.1% of the total weight of the sample (Figure 4.6).

Take into account that we can force the granules aggregate to pass through the sieve by applying hand pressure, but we can do it only for sieve size 20 mm and more.

Then, measure the weight of the remaining aggregate on the sieves separately and calculate the aggregate passing through the sieves as shown in Table 4.7.

The nominal maximum aggregate size is defined as the smallest sieve that passes at least 95% of the coarse aggregate or whole aggregate.

Do not put too much weight on the sieve as the maximum weight of the remaining aggregate on the sieve should not exceed the weights shown in Table 4.8, which are based on Egyptian standard specifications ESS:2421:1993 and BS 812: section 103.1:1985.

Table 4.9 is based on ISO 6274-1982. The acceptance and refusal limits for coarse aggregate and sand, and whole aggregate based on ECP codes, BS 882:1992, and ASTM C33-93, are shown in Tables 4.10–4.12. Table 4.13 shows the grading requirement of coarse aggregate according to BS 882:1992, which is similar to the ECP but with some modifications. The acceptable grade requirements in BS 882:1973 are shown in Table 4.14.

FIGURE 4.5   Shapes of sieves.

(a)

(b)

FIGURE 4.6   (a) Sand on sieves. (b) Washing aggregate.

TABLE 4.7

Calculating Percentage of Remaining Aggregate and Aggregate Passing through Sieve

Sieve Size (mm)	Remaining Weight on Sieve	Total Remaining Weight on Sieve	Percentage of Remaining Aggregate	Percentage of Passing Aggregate
37.5	W_1	W_1	$R_1 = W_1/W$	$100 - R_1$
20	W_2	$W_1 + W_2$	$R_2 = (W_1 + W_2)/W$	$100 - R_2$
10	W_3	$W_1 + W_2 + W_3$	$R_3 = (W_1 + W_2 + W_3)/W$	$100 - R_3$
5	W_4	$W_1 + W_2 + W_3 + W_4$	$R_4 = (W_1 + W_2 + W_3 + W_4)/W$	$100 - R_4$

TABLE 4.8

Maximum Weight for Remaining Aggregate in Different Sieve Sizes

Sieve Opening Size (mm)	Max. Weight (gm)		Sieve Opening Size (mm)	Max. Weight (gm)	
	Sieve Diameter 450 mm	Sieve Diameter 300 m		Sieve Diameter 300 m	Sieve Diameter 200 m
50	14	5	5.00	750	350
37.5	10	4	3.35	550	250
28	8	3	2.36	450	200
20	6	2.5	1.80	375	150
14	4	2	1.18	300	125
10	3	1.5	0.85	260	115
6.3	2	1	0.6	225	100
5	1.5	0.75	0.425	180	80
3.35	1	0.55	0.300	150	65
			0.212	130	60
			0.15	110	50
			0.075	75	30

TABLE 4.9

Acceptance and Refusal Limits for Fine Aggregate

	Percentage Passing Sieve				
	ECP and BS882:1992				
Sieve Opening Size (mm)	General Grading	Coarse	Medium	Fine	ASTM C33-93
10.0	100	—	—	—	100
5.0	100–89	—	—	—	95–100
2.36	100–60	100–60	100–65	100–80	80–100
1.18	100–30	90–30	100–45	100–75	50–85
0.6	100–15	45–15	80–25	100–55	25–60
0.3	70–5	40–5	48–5	70–5	10–30
0.15	15–0	—	—	—	2–10

TABLE 4.10

Acceptance and Refusal Limits for Coarse Aggregate in Egyptian Code

	Percentage by Weight Passing Sieve						
Sieve Opening Size (mm)	Nominal Size of Graded Aggregate (mm)			Nominal Size of Single-Sized Aggregate (mm)			
	5–40	5–20	5–10	40	20	14	10
50.0	100	—	—	100	—	—	—
37.5	100–90	100	—	100–85	100	—	—
20.0	70–35	100–90	100	25–0	100–85	100	—
14.0	—	—	100–90	—	—	100–85	—
10.0	40–10	60–30	85–50	5–0	25–0	50–0	100
5.0	5–0	10–0	10–0	—	5–0	10–0	100–50
2.36	—	—	—	—	—	—	30–0

TABLE 4.11

Acceptance and Refusal Limits for Coarse Aggregate in BS882:1992

	Percentage by Weight Passing Sieve						
	Nominal Size of Graded Aggregate (mm)			Nominal Size of Single-Sized Aggregate (mm)			
Sieve Opening Size (mm)	5–40	5–20	5–14	40	20	14	10
50.0	100	—	—	100	—	—	—
37.5	100–90	100	—	100–85	100	—	—
20.0	70–35	100–90	100	25–0	100–85	100	—
14.0	25–55	40–80	100–90	—	0–70	100–85	100
10.0	40–10	60–30	85–50	5–0	25–0	50–0	85–100
5.0	5–0	10–0	10–0	—	5–0	10–0	25–0
2.36	—	—	—	—	—	—	5–0

TABLE 4.12

Grading Requirements for Coarse Aggregate in ASTM C33-93

	Percentage by Weight Passing Sieve				
	Nominal Size of Graded Aggregate (mm)			Nominal Size of Single-Sized Aggregate (mm)	
Sieve Opening Size (mm)	37.5–4.75	19.0–4.75	12.5–4.75	63	37.5
75	—	—	—	100	—
63	—	—	—	90–100	—
50.0	100	—	—	35–70	100
38.1	95–100	—	—	0–15	90–100
25	—	100	—	—	20–55
19	35–70	90–100	100	0–5	0–15
12.5	—	—	90–100	—	—
9.5	10–30	20–55	40–70	—	0–5
4.75	0–5	0–10	0–15	—	—
2.36	—	0–5	0–5	—	—

TABLE 4.13
Acceptance and Refusal Limits for Whole Aggregate in BS882:1992 and ECP2002

	Percentage Passing Sieve		
Sieve Opening Size (mm)	Nominal Max. Aggregate Size 40 mm	Nominal Max. Aggregate Size 20 mm	Nominal Max. Aggregate Size 10 mm
50.0	100	—	—
37.5	100–95	100	—
20.0	80–45	100–95	—
14.0	—	—	100
10.0	—	—	100–95
5.0	50–25	55–35	65–30
2.36	—	—	50–20
1.18	—	—	40–1
0.60	30–8	10–35	30–10
0.30	—	—	15–5
0.15	8–0[a]	8–0[a]	8–0[a]

[a] Increase to 10% for crushed rock fine aggregate.

TABLE 4.14
Grading Requirement for Whole Aggregate Based on BS882:1973

	Percentage by Weight Passing Sieve	
Sieve Size (mm)	Nominal Size 40 mm	Nominal Size 20 mm
75	100	—
37.5	95–100	100
20	45–80	95–100
5	25–50	35–55
600 μm	8–30	10–35
150 μm	0–6	0–6

The grading requirements depend on the shape and surface characteristics of the particles. For instance, sharp, annular particles with rough surfaces should have a slightly finer grading to reduce the interlocking and to compensate for the high friction among the particles. The grading of crushed aggregate is affected primarily by the type of crushing plant employed. A roll granulator usually produces fewer fines than other types of crushers, but the grading depends also on the amount of material fed into the crusher.

FIGURE 4.7   Oven according to BS1377.

4.2.2.2   Abrasion Resistance of Coarse Aggregates in Los Angeles Test

From this test, one can define the abrasion factor, which is the percentage of lost weight due to abrasion.

The Los Angeles device is a cylinder that is rotated manually. It contains balls made of cast iron or steel with diameter of about 48 mm and weight per ball ranges between 3.82 and 4.36 Newton.

This test begins by taking a sample of big aggregate, with weight ranging from 5 to 10 kg, washing the sample with water and then drying it in an oven at 105°C–110°C as shown in Figure 4.7.

Separate the samples into different sizes through the sieves as shown in Table 4.15 and collect the test samples from the aggregate by mixing the weights from the table.

The sample will be weighed after remixing. The weight is W1 and the grading type is as per Table 4.15 from A to G. By knowing the grading type define the number of bars that will be put in the device as shown in Table 4.16.

Put the sample and the balls inside the Los Angeles machine and rotate at 10–31 rpm so that the total number of rotations is 500 for sample gradients A, B, C, D, F and 1000 cycles for the rest of the gradings.

Lift the aggregate from the machine and put it in sieve size 16 mm and then through sieve size 1.7 mm. Wash the aggregates that remain on the two sieves and then dry in oven at 105°C–110°C and then weigh it (W2).

$$\text{Percentage of abrasion} = W1 - W2/W \times 100$$

The acceptable percentage of abrasion based on the Los Angeles test is not more than 20% for aggregate and 30% for crushed stone.

TABLE 4.15
Weights after Sieve Analysis

Passing from	Remaining on	Grade						
		A	B	C	D	E	F	G
75.00	63.00					1500		
63.00	50.00					1500		
50.00	37.5					1500	5000	
37.5	25.00	1250					5000	5000
25.00	19.00	1250						5000
19.00	12.5	1150	1500					
12.5	9.5	1150	1500					
9.5	6.3			1500				
6.3	4.75			1500				
4.75	2.38				5000			

TABLE 4.16
Determining Number of Abrasion Ball

Grading Type	Number
A	13
B	11
C	8
D	6
E	11
F	11
G	11

4.2.2.3 Determination of Clay and Other Fine Materials in Aggregates

Clays and fines are known as soft material that passes through micron sieve size. This test ensures that aggregate is in conformity with the standard specifications and this will be done by taking a sample whose weight is at least 250 grams in the case of fine aggregate. In the case of coarse aggregate or whole aggregate the sample weight will be as in Table 4.17.

The test begins by drying the sample in an oven at 110°C ± 5°C until (W) and then immerse the sample in water and move it strongly to remove the clay and fine materials by putting the wash water on 75 micron and 141 micron sieves. Repeat the washing several times until the washing water is pure.

Return the retaining materials at 141 and 75 micron sieves to the washed sample and dry the remaining at the same oven temperature until it reaches W1 and then calculate the percentage of clay and fine materials by weight as follows:

TABLE 4.17

Sample Weight to Test Percentages of Clay and Fine Aggregate

Max. Nominal Aggregate Size (mm)	Least Sample Weight (kg)
4.75–9.5	5
9.5–19	15
19–37.5	25
<37.5	50

TABLE 4.18

Maximum Allowable Limits for Fine Aggregate in Concrete

Type of Aggregate	Percentage of Clay and Fine Materials by Weight (%)
Sand	3
Fine aggregate from crushing stone	5
Coarse aggregate	1
Coarse aggregate from crushed stone	3

Percentage of fine materials and clay = $(W - W1)/W \times 100$

Based on the ECP, ASTM C142-78, and BS882:1992 specifications, the maximum acceptable limits for clay and fine materials in aggregate are shown in Table 4.18.

4.2.2.3.1 On-Site Test

This simple test can be conducted at the site. Add 50 cm³ of clean water to fill a glass tube and then add sand until the total volume is 100 cm³. Add clean water until the total volume is 150 cm³ as shown in Figure 4.8.

The glass tube is shaken strongly until the fine materials and clay move to the top. The tube will be set in a horizontal table for 3 hours. Calculate the percentage of fine materials and clay by volume on the top layer of the water with respect to the volume of the aggregate settling in the bottom of the tube.

If you find a higher percentage of suspended clay and fine materials, the aggregate does not match with the specification, so you should review the previous laboratory test which is the official test to accept or refuse the aggregate.

4.2.2.4 Aggregate Specific Gravity Test

The specific gravity of the aggregate is apparent density relative to the rule of solid constituent and there are no spaces inside it to access the water. The specific gravity is obtained by dividing the weight of dry aggregate by water volume equivalent to its volume (Table 4.19).

FIGURE 4.8   Tubes for aggregate density test.

TABLE 4.19
Aggregate Specific Gravities

Type of Aggregate	Range of Specific Gravity
Sand	2.5–2.75
Coarse aggregate	2.5–2.75
Granite	2.6–2.8
Basalt	2.6–2.8
Limestone	2.6–2.8

4.2.2.5   Fine Aggregate Test

The test sample is not to exceed 100 g and is dried in the oven at 100°C to 110°C, cooled in the dryer, weighed and returned to the drying process, and then weighed until reaching proved weight and measurement (W).

Add water at 15°C to 25°C to a graduated tube and then add the fine aggregate (W) and leave it submerged in the water for an hour after removing the bubbles by knocking the tube.

One hour after adding the aggregates read the level of the water in the tube then define the aggregate volume (V), then calculate the specific gravity by the following equation:

$$\text{Specific gravity of aggregate} = W/V$$

4.2.2.6   Define Specific Gravity for Coarse Aggregate

Immerse a sample of about 2 kg in water at 15°C to 12°C for 24 hours. Then take the aggregate and dry it manually with a piece of wool. Put a defined volume of water in a big bowl, whose volume is known (V1).

TABLE 4.20

Sizes of Containers to Define Aggregate Bulk Density

Max. Nominal Aggregate Size (mm)	Bowl Volume (L)	Bowl Dimensions (mm)		
		Internal Diameter	Internal Height	Thickness
>40	30	360	293.6	5.4
40–5	15	360	282.4	4.1
<5	3	155	158.9	3.0

Add the aggregate to the bowl to fill to its midpoint, then add water until it fills the bowl completely, which identifies volume (V2), and then take the aggregate, dry it, and measure its weight (W). Calculate the coarse aggregate specific gravity using the following equation:

$$\text{Corase aggregate specific gravity (Sg)} = W/(V2 - V1)$$

4.2.2.7 Bulk Density or Volumetric Weight Test for Aggregate

This test determines volumetric weight of the aggregate. By knowing the volumetric weight you can transform a given volume of aggregate so it is equivalent to the weight or vice versa. By knowing the volumetric weight and specific gravity you can calculate the percentage of voids between aggregate grains.

Volumetric weight is the ratio of the weight of the aggregate to the volume that it occupies. The percentage of voids is the ratio between voids of the aggregate and the total volume of the aggregate that it occupies. Define container size based on Table 4.20.

By knowing the volume of the bowl (V), measure the bowl weight when it is empty (W1). Then fill the contaner with the aggregate and perform standard compaction by standard rod for 25 times. Do that twice until the bowl is completely full. Then measure the weight of the bowl with the compacted aggregate (W2).

The volumetric weight can be calculated using the following equations:

$$\text{Volumetric weight } (V_w) = (W2 - W1)/V$$

$$\text{Void percentage} = (V_w - S_g)/V_w \times 100$$

4.2.2.8 Percentage of Aggregate Absorption

This test is used to identify the absorption of water by the aggregate with maximum nominal size higher than 5 mm, and Table 4.21 shows the limits of absorbing aggregate that should be followed in selecting the aggregate as it will affect the percentage of water in the concrete mix.

Determine the weight about 100 times the maximum nominal aggregate size in millimeters. Wash the sample before the test on a sieve of 5 mm to remove the suspended materials.

TABLE 4.21
Maximum Allowable Limit of Absorption of
Water by Aggregate

Type of Aggregate	Percentage of Absorption (%)
Quartized and crushed limestone	1–0.5
Granite	0–1
Stone	Not more than 2.5

Put the sample in a wire mesh of 1–3 mm and then immerse it in a container full of water at a constant temperature of 15°C to 25°C so that the total immersion is less than the distance between the highest point in the basket and surface water on 50 mm. After immersion remove the entrained air and leave the basket and sample immersing for 24 hours.

Remove the basket and the sample from the water. Dry the surfaces of the sample gently and distribute it in a piece of cloth. Leave it in the air and away from the sun or any source of heat and then weigh it (W1).

Then put the sample in an oven at $105 \pm 5°C$ for 24 hours, let it cool without being exposed to any humidity and then weigh it (W2).

$$\text{Absorption percentage} = (W1 - W2)/W2 \times 100$$

4.2.2.9 Recycled Aggregate Concrete

The idea of recycling of aggregates was introduced many years ago to solve the growing waste disposal crisis and protect depleted natural sources as stated by Kasai in 1988. Good natural aggregates are very expensive in many regions.

In the past the strength of aggregates derived from concrete structures was relatively low and consequently the applications were of secondary importance. At present demolition of RC or PC structures made from strong concrete, like building frames, bridge beams, airport runways, or rejected precast members in plants creates sources of recycled aggregates that have different qualities.

Presently such a situation is typical for countries in Central and Eastern Europe where the programs of modernization have started for roads, bridges, and municipal and industrial structures. Sometimes it is necessary to demolish relatively new structures because the functional properties do not fit new projects. The best examples are road bridges for which the prefabricated prestressed concrete beams with spans 15 to 18 meters are no longer sufficient and have to be removed to widen the spans of structures that overpass highways.

Concrete recycling presents a new economical aspect. Concrete originally mixed with a large amount of cement retains some binding abilities, particularly when the carbonated zone is not too deep. It may be activated with silica fume or fly ash admixtures. Some savings in cement may be obtained in this way, as studied by Salem and Burdette (1998).

A pilot material test was undertaken to explain how to obtain good quality structural concrete using aggregates from demolished structures made from formerly medium- or high-strength concrete, and to determine what properties could be obtained by introduction of silica fume and superplasticizers. Basic results were presented by Ajdukiewicz and Kliszczewicz in 2000 during the PCI/FHWA/FIB Symposium in Orlando.

Many tests of recycled aggregates as the components of structural concrete have been undertaken since early 1980s. Despite serious differences in the original formulations, the general conclusions are that recycled aggregate should be considered as a valuable material. Such conclusions are also valid for high-strength, high-performance recycled concrete.

Nevertheless, when properties of concrete with recycled aggregates are compared with properties of corresponding concrete with natural (new) aggregates the following differences have been noticed:

- Lower compressive strength from 10% to 30%
- Slightly lower tensile strength not more than 10%
- Lower elasticity modulus from 10% to 40% (depending on origin of coarse aggregate)
- Substantially greater shrinkage from 20% to 55% but slightly smaller creep up to 10%

No significant differences were observed in bond (tested by RILEM method) and in freezing resistance. Some changes in properties of fresh mix of concrete were recorded: shorter setting time and faster decrease of workability.

Results of tests presented in different countries have been mainly concerned with properties obtained from testing concrete specimens prepared with various recycled aggregates. Mukai and Kikuchi (1988) and Di Niro et al. (1998) performed tests on structural members, particularly those made from high-strength concrete. Such tests are necessary because it is difficult to predict the influence of the combination of properties on the overall behavior of reinforced concrete members made from recycled aggregate concrete

According to tests carried out by Ajdukiewicz and Kliszczewicz (2002) for four series of beams with two concrete grades about 40 to 90 MPa, behavior of reinforced concrete beams subjected to bending and shear depended on many properties of materials and constructional features. Taking into account similar elements made from concrete of the same mixture proportions but using different aggregates it is difficult to predict their failure, shape, load-bearing capacity, deflection, etc. Such a situation was found with the introduction of recycled aggregates. Despite complete tests on differences of properties of recycled aggregates, the assessment of global results may be based only on standard methods of analysis.

Comparison of observations and results of tests made on four series of beams are as follows:

- Differences in behavior of beams were relatively small within a particular series.

TABLE 4.22
Maximum Allowable Limits of Salt and Suspended Materials in Water

Type of Salt and Suspended Materials	Max. Limit of Salt Content (g/L)
Total dissolved solid (TDS)	2.0
Chloride salt	0.5
Sulfate salt	0.3
Carbonate and bicarbonate salt	1.0
Sodium sulfate	0.1
Organic materials	0.2
Clay and suspended materials	3.0

- Load-bearing capacity of beams made from concretes with various amounts of recycled aggregate differed relatively more with medium-strength concrete (30–60 MPa) than with high-strength concrete (80–90 MPa).
- Deflections (immediate) of beams made from recycled aggregate concrete were always greater than in comparable beams made with natural (new) aggregate, but the range of differences varied from about 10% to 25% at failure load and as much as 30% to 50% at a probable service load.

4.2.3 MIXING WATER TEST

Water is a very essential factor for mixing concrete and for the curing process, so the quality of water is essential to concrete durability as it must follow the project specifications. Table 4.22 illustrates the acceptable water specifications for the mixing and curing process.

The determination of soluble salts in water can be obtained by taking a sample of approximately 25 mL of water placed in a dish of platinum. Evaporate the sample and then transfer it to the drying oven at 105°C until you fix the measured weight. By knowing the size of the sample, you can get the total content of dissolved salts in water (TDS) (g/L).

Chloride content in water is defined by performing a chemical test to compare the content of chloride with the permissible content in the specifications as defined in Table 4.22.

4.3 STEEL REINFORCEMENT TEST

The steel reinforcement is an important element of reinforced concrete structures where the tensile loads are met by reinforcing steel bars in concrete; therefore, we must make sure that rebar meets specifications for the project strictly. Therefore, we will present here some important tests for quality control of rebar.

TABLE 4.23

Weight of Deformed Bar per Unit Length

Nominal Diameter (mm)	Nominal Cross Sectional Area (mm²)	Weight (kg/m')	Tolerance Allowance (%)
6	28.3	0.222	±8
8	50.3	0.395	
10	78.5	0.618	±5
12	113	0.888	
13	133	1.04	
14	154	1.21	
16	201	1.58	
18	254	2.00	
19	283	2.22	
20	314	2.47	
22	380	2.98	
25	461	3.85	±4
28	616	4.83	
32	804	6.31	
36	1020	7.99	
40	1257	9.86	
50	1964	15.41	

4.3.1 WEIGHTS AND MEASUREMENT TEST

This test requires a delicate balance, measuring tape, and relevant Vernier.

Withdraw two samples of the same diameter of each consignment weighing less than 50 tons. Withdraw three samples if the consignment weighs more than 50 tons.

To determine bar diameter, in each sample measure two perpendicular diameters in the same cross section by using a special measurement unit. Ascertaining the weight of longitudinal meter is performed by weighing a sample with length not less than 500 mm with accuracy ± 0.5% (Table 4.23).

The actual cross sectional area, A, is calculated by taking into account the density of steel, 7.85 ton/m³, the weight to the nearest gram, and L (length):

$$\text{Actual cross sectional area} = W/(0.00785\ L)$$

$$W = \text{weight to nearest kg}$$

$$L = \text{length to nearest mm}$$

4.3.2 TENSION TEST

This test is useful to define the mechanical properties of steel bars by exposing a test sample with the standard dimensions to tensile stress until fracture. Figure 4.9

FIGURE 4.9   Steel bar in tension machine.

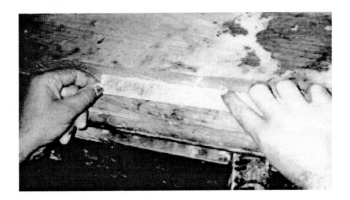

FIGURE 4.10   Elongation measurement.

presents the shape of the tension machine, and Figure 4.10 shows the elongation measurement. From this test one can define the yield stress or proof stress and also the elongation percentage.

$$\text{Elongation percentage} = (L_2 - L_1)/L_1 \times 100$$

where
  L_1 is the original measured length.
  L_2 is the final measured length.

The length of the short sample $L_1 = 5D$ and the length of the long sample $L_1 = 10D$, where D is the diameter of the steel bar, and the deviation for the measurement of sample dimensions are not more than ±0.5%.

TABLE 4.24
Minimum Acceptance Limits for Mechanical Properties

Steel Grade	Min. Yield Strength (N/mm²)	Min. Tensile Strength (N/mm²)	Elongation (%)
240	240	350	20
280	280	450	18
360	360	520	12
400	400	600	10

During sample preparation you can do some modification, but you cannot change the shape of the sample by increasing its temperature such as exposing the sample to heat.

$$\text{Yield strength} = \text{yield force/actual cross sectional area}$$

$$\text{Tension strength} = \text{maximum force/actual cross sectional area}$$

The Egyptian standard requires at least 95% of the quantity-tested values set in Table 4.24. Moreover, the result of any single test is not less than 95% of the values mentioned in Table 4.24. It can be agreed between the supplier and customer to ensure that the values shown in Table 4.24 are the minimum acceptance limits to the bars.

The ratio between the tension strength and yield strength for any sample is not less than 1.1 and 1.05 for smooth and ripped bar, respectively. For steel for which the yield point cannot be clearly defined, define the proof stress 0.02% instead of yield stress.

4.4 NONDESTRUCTIVE TEST FOR HARDENED CONCRETE

We may need to conduct some tests and measurements to ascertain the quality of hardened concrete or resolve a disagreement between the contractor and the owner or the consultant, or deal with lack of confidence in a material.

In that case, the quality control inspector should be able to deal with different ways to detect the quality of hardened concrete through nondestructive testing.

Therefore, choose the appropriate method that is economically feasible as well as from a structure condition and system point of view. The tests should be based on the condition of the structure and the location of the member that will be tested to be structurally safe, and the test should provide reasonable confidence and accuracy.

4.4.1 CORE TEST

This test is semidestructive and is very important for the study of the safety of structure as a result of changing the system of loading or a deterioration as a result of fire

TABLE 4.25

Number of Cores and Deviation in Strength

Number of Cores	Deviation Limit between Expected Strength and Actual Strength (Confidence Level 95%)
1	+12%
2	+6%
3	+4%
4	+3%

or weather factors, or the need for temporary support for repair in the absence of accurate data about concrete strength. This test is not too expensive and it is the most accurate for finding the strength of concrete actually carried out.

Core testing is done by cutting cylinders from concrete members and the cutting may affect the integrity of the structure. The samples must be taken in strict adherence to the standard. This will ensure accuracy of the results without weakening the structure. Deterioration caused by corrosion of steel bars causes a structure to lose most of its strength. This situation requires more caution when performing core testing by selecting an appropriate concrete member to prevent structural damage.

The codes and specifications provide some guidance to the number of cores to test and these values are as follows:

- Volume of concrete member (V) ≤ 150 m³—three cores
- Volume of concrete member (V) > 150 m³—[3 + (V − 150/50)] cores

The degree of confidence of the core test depends on the number of tests, which must be minimal. The relation between number of cores and confidence is shown in Table 4.25.

Before you choose the location of the sample, first define the location of the steel bars to assist you in selecting a sample away from the steel bars to avoid the possibility of taking a sample containing steel reinforcement bars.

We must preserve the integrity of the structure, and therefore this test should be performed by an experienced engineer. When conducting such an experiment one should take precautions, determine the responsibilities of individuals, and ensure that the nondestructive testing has been conducted in a correct manner. Figures 4.11 and 4.12 present the process of taking the core from a reinforced concrete bridge girder.

4.4.1.1 Core Size

Note that the permitted diameter is 100 mm in the case of maximum aggregate size 25 mm and 150 mm if the maximum aggregate size does not exceed 40 mm. It is preferable to use 150 mm diameter whenever possible as it gives more accurate results, as shown in Table 4.26, which shows the relationship between the dimensions of the sample and potential problems. This table should be discussed when choosing the reasonable core size.

FIGURE 4.11   Core sampling.

FIGURE 4.12   Core sampling. Note reinforced girder.

Some studies stated that the test can be done with core diameter 50 mm in the case of maximum aggregate size and not more than 20 mm overall; small cores sizes give different results than large sizes.

Because of the seriousness of the test and the inability to take high numbers of samples, good supervision is needed when taking the sample. Moreover, the laboratory test must be certified and the test equipment must be calibrated with a certificate from a certified company.

Sample extraction is done using pieces of a cylinder that is a different country-cylinders equipped with ransom of special alloy mixture with diamond powder to feature pieces in the concrete during the rotation of the cylinder through the body. Precautions sampling method should be a match and be consistent pressure to bear on the match by appropriate and it depends on the expertise of the technician.

TABLE 4.26
Core Sizes and Possible Problems

Test	Diameter (mm)	Length (mm)	Possible Problem
One	150	150	May contain steel reinforcement
Two	150	300	May cause more cutting depth to concrete member
Three	100	100	Not allowed if maximum aggregate size is 25 mm
			May cut with depth less than required
Last	100	200	Less accurate data

After that the core will be filled with dry concrete of suitable strength or grouting. Another solution is to use epoxy, which is injected in the hole, and then insert a concrete core of the same size to close the hole.

Regardless of the method used, the filling must be done soon after cutting. The filling material must be ready for use by the technician who does the cutting as the core may affect the integrity of the structure.

The lab must examine and photograph every core, and note gaps within the core as small voids if they measure between 0.5 and 3 mm, or voids if the average measurement is between 3 and 6 mm, or big voids if more than 6 mm. This test also examines nesting and determines the shape, kind, and color gradients of aggregates and any apparent qualities of the of sands.

In the laboratory the dimensions, weight of each core, the density, steel bar diameter, and distance between the bars will be measured.

4.4.1.2 Sample Preparation for Test

After cutting the core from the concrete element, process the sample for testing by leveling the surface of the core, then take the core, which has a length not less than 95% of the diameter and not more than double the diameter.

Figure 4.13 shows the shape of the core sampling after cutting from the concrete structure member directly.

Level the surface using a chainsaw concrete or steel cutting disk. After that prepare the two ends of the sample by covering them with mortar or sulfide and submerge the sample in water at 20°C ± 2°C for at least 48 hours before testing the sample.

The sample is put in a machine and influence load is applied gradually with the rate of regular and continuous range of 0.2 to 0.4 N/mm^2 until it reaches the maximum load at which the sample is crushed.

Determine the estimated actual strength for cube by knowing the crushing stress, which is obtained from the test using the following equation, as λ is the divided core length to its diameter. In the case of horizontal core the strength calculation will be as follows:

$$\text{Estimated actual strength for cube} = 2.5/(1/\lambda) + 1.5 \times \text{core strength}$$

where λ = core length/core diameter.

FIGURE 4.13 Shape of core.

In the case of vertical core the strength calculation will be as follows:

$$\text{Estimated actual strength for cube} = 2.3/(1/\lambda) + 1.5 \times \text{core strength} \qquad (4.1)$$

In the case of existing steel in the core perpendicular to the core axis the previous equations will be multiplied by the following correction factor:

$$\text{Correction factor} = 1 + 1.5 \, (s \, \phi)/(LD) \qquad (4.2)$$

where
  L = the core length.
  D = core diameter.
  S = distance from steel bar to edge of core.
  ϕ = steel bar diameter.

Cores are preferred to be free of steel, but if steel is found you must use the correction factor only in the event that the value ranges from 10%–5%. If the correction

TABLE 4.27
Comparison of Standard Cylinders and Cores

Age (days)	Standard Cylinder Strength (MPa)	Core Strength (MPa)	fc (core)/fc (cylinder at 28)
7	66	57.9	0.72
28	80.4	58.5	0.73
56	86.0	61.2	0.76
180	97.9	70.6	0.88
365	101.3	75.4	0.94

factor exceeds 10%, the results of the cores cannot be trusted and you should take another core.

The cores are often taken after 28 days. But practically speaking, the situation was different, as shown in Table 4.27, in a study performed by Yuan et al. in 1991 for reevaluation of core strength for high-strength concrete by comparing the standard cylinders and cores taken from a column cured using a sealing compound. It can be seen that in situ concrete often gains little strength after 28 days.

When examining the test results, the following points must be taken into account:

- Before the test submerge the sample in water as this decreases strength up to about 15% for dry concrete.
- The equation to calculate the expected concrete strength does not take into account any differences in direction between the core and the standard cube direction.
- The concrete is acceptable if the average strength to the cores is at least 75% of the required strength and the calculated strength for any core is less than 65% of the required strength.
- In the case of prestressed concrete, the strength is acceptable if the average strength to the cores is at least 80% of the required strength and the calculated strength for any core is less than 75% of the required strength.

4.4.2 REBOUND HAMMER

This is nondestructive testing for determining the estimated concrete compressive strength. This is the most common method as it is easy and cheap compared with other tests, but gives less precise results.

This test relies on measuring the concrete strength by measuring the hardening from the surface. It is used to identify the concrete compressive strength of the concrete member by using calibration curves of the relationship between hardening and compressive strength. Figures 4.14 and 4.15 present different types of rebound (Schmidt) hammers. Most commonly they give impact energy of 2.2 N/mm. The reading will be analog or digital or connect to a memory card to record the readings. Figures 4.15–4.17 show Schmidt (rebound) hammers and reading the results technique.

FIGURE 4.14 Rebound hammer.

FIGURE 4.15 Rebound hammer.

FIGURE 4.16 Testing with rebound hammer.

FIGURE 4.17   Reading results.

Inspect the device before using the calibration tools included with the device. The calibration should be within the allowable limit based on manufacturer recommendations.

The first and most important step in the test is to clean and smooth the concrete surface at the sites that will be tested by measuring an area about 300 × 300 mm. Preferably test on a surface that has no change after casting or a surface that has not had any smoothing during the casting process.

On the surface to be tested draw a net of perpendicular lines in both directions that are 2 to 5 cm apart. The intersection points will be tested and the test point in any case must be away from the edge by about 2 cm.

As shown in Figures 4.16 and 4.17, the surface must be cleaned before the test and the rebound hammer is perpendicular to the surface. The following recommendations apply:

- The hammer must be perpendicular to the surface that will be tested because the direction of the hammer affects the value of rebound as a result of the impact of hammer weight.
- A wet surface gives significantly lower readings of the rebound hammer than a dry surface by up to 20%.
- The tested concrete member must be fixed so it does not vibrate.
- Do not use the curves for the relationship between concrete compressive strength and rebound number as given by the manufacturer. Calibrate the hammer by taking the reading on concrete cubes and crushing the concrete cubes to obtain the calibration of the curves. This calibration is important from time to time as the spring inside the rebound hammer loses some of its stiffness with time.
- You must use one hammer only when comparing quality of concrete at different sites.
- The type of cement affects the readings as in the case of concrete with high-alumina cement, which can yield higher results than concrete made with ordinary Portland cement by about 100%.

- Concrete made with sulfate-resistant cement can yield results of about 50% less than ordinary Portland cement.
- Higher cement content gives lower readings than concrete with less cement content; in any case the gross error is only 10%.

4.4.2.1   Data Analysis

The number of readings must be high enough to give reasonably accurate results. The minimum number of readings is 10, but usually we take 15 readings.

The extreme values will be excluded and the other remaining values are averaged. From this the concrete compressive strength will be known and the results will be compared with the required concrete strength.

4.4.3   Ultrasonic Pulse Velocity

This is a nondestructive test. The concept is to measure the speed of transmission of ultrasonic pulses through the construction member. By measuring the time required for the transmission of impulses and by knowing the distance between the sender and receiver the pulse velocity can be calculated.

The calibration of these velocities is done by finding the concrete strength and its mechanical characteristics. You can use the same procedure to identify compressive strength, dynamic and static modulus of elasticity, and Poisson ratio. The equipment must have the capability to record time for the tracks with lengths ranging from 100 to 3000 mm accurately + 1%.

The manufacturer should define the methods of working of the equipment and different temperature and humidity requirements. The device must have a power transformer sender and receiver of natural frequency vibrations between 20 to 150 kHz, bearing in mind that the frequency appropriate for most practical applications of concrete is 50–60 kHz.

Surface transmission is illustrated in Figure 4.18a and b. Semi–direct transmission is shown in Figure 4.19a and b, and wave transmission, which is a direct transmission, is shown in Figure 4.20a and b.

Provided with the UT equipment are two rods of metal with lengths of 250 mm and 1000 mm. The first is used in the determination of zero of the measurement, and the second is used in the calibration. Both rods show the time of the passage of waves. Hence, connect the ends of the rods in an appropriate way and measure the time for pulse transmission, and compare it with the known reading. For the small rod if there are any deviations, adjust the zero of the equipment to show the known reading.

The long bar is used in the same way to define the accuracy of the result. In this case the difference between the two readings should be not more than ± 0.5% to achieve the required accuracy.

The wave transmission velocity value in steel is twice the value in concrete, so in case of steel bars in concrete members, testing will influence the accuracy of the reading as it will be high for the wave impulse velocity and to avoid that the location of the steel reinforcement must be defined previously with respect to the path of the ultrasonic pulse velocity.

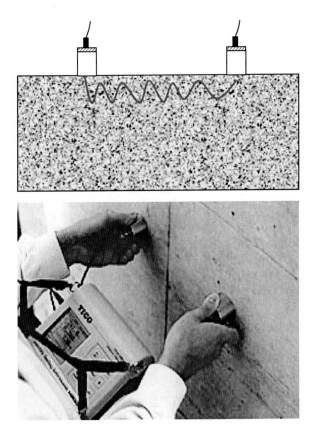

FIGURE 4.18   Surface transmission.

To get a correct reading, the possibility of steel bars being parallel to the path of the pulse wave, as shown in Figure 4.21, must be considered. The calculation of the pulse velocity will be as shown in the following equation. By knowing the wave path, the wave velocity in steel, and the distance between the two ends time of transmission will be calculated by the equipment.

$$V_C = K \cdot V_m \tag{4.3}$$

$$K = \gamma + 2(a/L)(1 - \gamma^2)^{0.5} \tag{4.4}$$

$$L_s = (L - 2b) \tag{4.5}$$

where
   V_m = pulse velocity from transmission time from the equipment.
   V_c = the pulse velocity in concrete.
   γ = a factor whose value varies according to steel bar diameter.

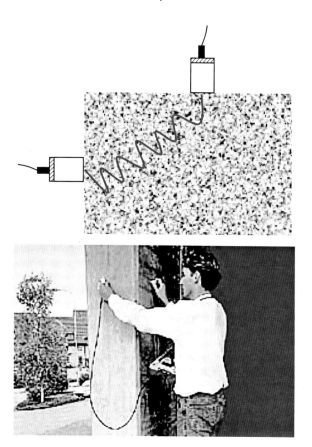

FIGURE 4.19   Semi-direct transmission.

The effect of the steel bar can be ignored if the diameter is 6 mm or less or if the distance between the steel bar and the end of the equipment is far.

If the steel reinforcement bar axis is perpendicular to the direction of pulse transmission, as shown in Figure 4.22, the effect of the steel on the reading will be less. The effect can be considered zero if we use transmission source of 54 kHz and the steel bar has a diameter less than 20 mm. The above equation is used if the frequency is less and the diameter is higher than 20 mm and by change of the value of γ according to the bar diameter by the data delivered by the equipment (Figure 4.23).

There are other factors that influence the measurement such as the temperature (Table 4.28), thawing, and concrete humidity and these effects must be considered. The most common errors include

1. Ignore using the reference bar to adjust zero, which will impact the accuracy of the results.
2. The concrete surface that is well leveled and smoothed after pouring may have properties different from the concrete in the core of the member and

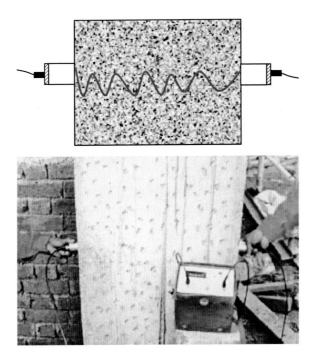

FIGURE 4.20 Direct transmission.

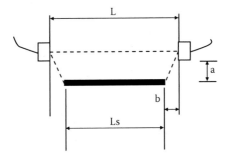

FIGURE 4.21 UT wave parallel to steel bars.

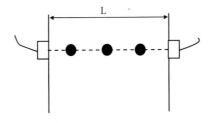

FIGURE 4.22 Wave of UT perpendicular to steel bars.

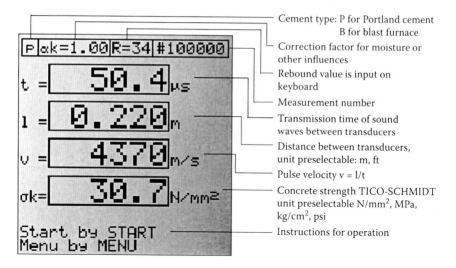

FIGURE 4.23 Ultrasonic pulse velocity screen.

therefore avoid it as much as possible in the measurements. If you cannot avoid that, you must take into account the impact of the surface.

3. Temperature affects the transmission ultrasonic velocity according to Table 4.28, which must be taken into account with increases or decreases in temperature of 30°C.

4. When comparing the quality of concrete of various components of the same structure similar circumstances should be taken into account in all cases in terms of the composition of concrete and moisture content and age, temperature, and type of equipment used. There is a relationship between the quality of concrete and the speed oriented, as shown in Table 4.29.

The static and dynamic moduli of elasticity can be defined by knowing the transmission pulse velocity in the concrete as shown in Table 4.30.

TABLE 4.28
Temperature Effect on Pulse
Transmission Velocity

Temperature (°C)	Correction of Velocity Reading (%)	
60	+5	+4
40	+2	+1.7
20	0	0
0	−0.5	−1
−4	−1.5	−7.5

TABLE 4.29
Relation of Concrete Quality and Pulse Velocity

Pulse Velocity (km/s)	Concrete Quality
>4.5	Excellent
4.5–3.5	Good
3.5–3.0	Fair
3.0–2.0	Poor
<2.0	Very poor

TABLE 4.30
Relation of Elastic Modulus and Pulse Velocity

Transmission Pulse Velocity (km/s)	Elastic Modulus (Mega N/mm²)	
	Dynamic	Static
3.6	24,000	13,000
3.8	26,000	15,000
4.0	29,000	18,000
4.2	32,000	22,000
4.4	36,000	27,000
4.6	42,000	34,000
4.8	49,000	43,000
5.0	58,000	52,000

4.4.4 LOAD TEST FOR CONCRETE MEMBERS

This test is done in the following conditions:

- If the core test gives results of concrete compressive strength lower than characteristic concrete strength, which is defined in design
- If this test is included in the project specifications
- If there is a doubt in the ability of the concrete structure member to withstand design loads

This test is usually done to the slabs and, in some cases, at the beams. The intent of this test is to expose the concrete slab to a certain load and then remove the load and measure the deformation on the concrete member. These deformations appear as deflection or cracks and are compared with the allowable limit in the specifications.

4.4.4.1 Test Procedure

This is done by loading the concrete member with load equal to the following:

$$\text{Load} = 0.85 \ (1.4 \text{ dead load} + 1.6 \text{ live load}) \tag{4.6}$$

The load is applied by using sacks of sand or concrete blocks. Sand sacks are calibrated at 10 sacks at least for every span of about 15 m^2 through direct weight of the sacks. Choose these sacks randomly to be representative and to determine the weight of an average sack. Then put these sacks on the concrete member that will be tested and take into consideration that there should be a distance between vertical sacks to prevent arch effect.

As for the concrete blocks, measure the weight and calibration and also take into account the horizontal distance left to avoid influencing the arch effect.

It is important to also identify the adjacent elements that have an impact on the structure element to be loaded in order to obtain the maximum possible deformation for the test member.

You must take into account, before loading processing, the location of the test by identifying the places where you will put the gauges, as well as calculating the actual dead load on the concrete member through the identification weight of the same member in addition to coverage such as tiles, which will be installed on a slab of concrete, as well as lower coverage and include plastering or weight of any kind of finishing work.

The location of the measurement unit is shown for a slab test as an example in Figure 4.24 and illustrates the following specifications:

1. One is placed in the middle of the span and another is placed beside it as a reserve as shown in Figure 4.24.
2. Put another measurement device at quarter the span from the support and then the consultant engineer must define the other reasonable location for measurement.
3. The measurement devices must be calibrated before use and preferably sensitivity below 0.01 mm and scale about 50 mm.
4. Use a device that measures the width of cracks and this device must have an accuracy of 0.01 mm.

4.4.4.2 Test Steps

- Define load test = 0.85(1.4 dead load + 1.6 live load) – dead load that already affects the member.
- Take the reading of the deflection before starting the test (R1).
- Start to put 25% of the test load and avoid the arch effect or any impact load.
- Read the measurement for the effect of 25% of the load and visually inspect the member to see if there are any cracks. If cracks are present measure their width.
- Repeat this procedure three times and for each time increase 25% of the load.

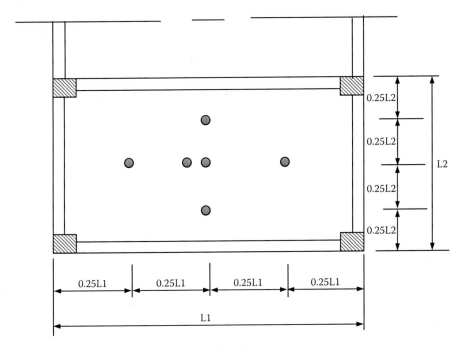

FIGURE 4.24   Location of measurement devices.

- Record the time for putting on the last load and the last deflection reading and crack thickness.
- After 24 hours of load affect time, record and draw the locations of the cracks and the maximum thickness and the deflection reading (R2) and then remove the load gradually and avoid any impact load.
- After removing all the load measure the deflection reading and crack width.
- Twenty-four hours after removing the load record the measurement and the reading (R3) and record the crack width.

4.4.4.3   Results Calculations

1. The maximum deflection after 24 hours from load effect:

$$\text{Maximum deflection} = (\text{first measurement after passing 24 hours} \\ \text{from load effect} - \text{reading before load effect}) \times \text{device sensitivity} \quad (4.7)$$

$$\text{Maximum deflection} = (R2 - R1) \times \text{device sensitivity} \quad (4.8)$$

If there is any problem in the first device use the second device reading, and if the readings of the two devices are close take the average of the two readings.

2. The remaining maximum deflection after 24 hours of completely removing the load will be from the following equation:

Maximum remaining deflection = (reading of the device after 24 hours from removing the load – reading before load effect) × device sensitivity   (4.9)

Maximum remaining deflection = (R3 – R2) × device sensitivity   (4.10)

3. The maximum recovery deflection will be calculated as follows:

The maximum deflection recovery = maximum deflection – maximum remaining deflection   (4.11)

The maximum deflection after 24 hours from loading and the recovery maximum deflection are as shown in Figure 4.24.

4. Draw the relation between the load and maximum deflection in the case of loading and uploading.
5. Note maximum crack thickness after 24 hours from load effect and 24 hours after removing the load.

4.4.4.4   Acceptance and Refusal Limits

• Calculate the maximum allowable deflection for the member as follows:

$$\text{Maximum allowable deflection} = L^2/2t, \text{ cm} \qquad (4.12)$$

where

    L = the span of the member in meters. The shorter span in the case of a flat slab and short direction for solid slab and cantilever will be twice the distance from the end of cantilever and the support face.

    t = thickness of the concrete member in centimeters.

• Compare the maximum deflection recorded after 24 hours from load effect and the allowable maximum deflection.
  • If the maximum deflection after 24 hours from load effect is less than the allowable maximum deflection from the previous equation then the test is successful and the member can carry load safely.
  • If the maximum deflection after 24 hours from load effect is higher than the allowable maximum deflection, the recover deflection after 24 hours from removing load must be equal or higher than 75% of maximum deflection.

Recovery deflection ≥ 0.75 maximum deflection

If this condition is verified, the member is considered adequate.
  • If the recovery is less than 75% of the maximum deflection, repeat the test by the same procedure but after 72 hours from removing the load from the first test. After repeating the test by the same procedure and precautions, this concrete structure member will be refused if not verified by the following two conditions:

1. If the recover deflection in the second test is less than 75% of the maximum deflection after 24 hours from load effect in second test
2. If the recorded maximum crack thickness is not allowed

4.4.5 PULLOUT TEST

This nondestructive test is is widely used in European countries and is called the lok-test in Denmark.

This test measures the concrete strength by means of a special jack, and measures the force required to pull out a previously cast-in metal insert with an enlarged end, as shown in Figure 4.25. The pull out force measured is related to the compressive strength of the concrete (Figure 4.26).

The test method is prescribed by ASTM C900-87 (reapproved 1993) and by BS 1881:207:1992. The ASTM states that the depth of the concrete above the enlarged end of the inserted must be equal to the diameter of the enlarged end. It also limits the apex angle of the frustum of the cone between 54 and 70 degrees.

4.4.6 DEFINE CHLORIDE CONTENT IN HARDENED CONCRETE

Determination of the maximum chloride content in concrete is an important factor in preventing the corrosion of the steel bars. Table 4.31 shows the limits of standardization of the content of dissolved chloride ions in the concrete.

The American code makes clear the limits of different content in chlorides and there are limits as in ACI 318R-89, and chloride ion limits, as stated in ACI 201,357,222, as a result of chlorides penetrating reinforced concrete. ACI 357 recommends that the water to be used in the concrete mixture should not contain more than 0.07% chloride in the case of reinforced concrete, or 0.04% in the case of pre-stressed concrete (Table 4.32).

European code of 1992 (ENV206) stated the limits of chlorides in concrete and identified these limits according to the type of application. These limits are 0.01%

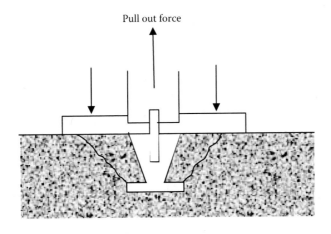

FIGURE 4.25   Pullout test.

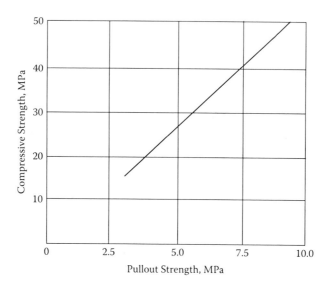

FIGURE 4.26   Relation between pullout strength and concrete compressive strength.

TABLE 4.31
Maximum Allowable Dissolved Chloride Ions

Condition around Concrete	Maximum Dissolved Chloride Ions in Concrete Water (% from Cement Weight)
Reinforced concrete exposed to chloride	0.15
Dry reinforced concrete totally protected from humidity	1.0
Different structure members	0.3

TABLE 4.32
ACI Recommendation for Maximum Acceptable Chloride Ions

Type of member	Dissolved (in water)[d]	Total[c]	Dissolved (in acid)[b]	Total[a]
Prestress concrete	0.06		0.06	0.08
Reinforced concrete exposed to chloride	0.15	0.1	0.1	0.2
Dry reinforced concrete	1.0			
Different structure members	0.3	0.15		

[a] ACI Committee 222
[b] ACI Committee 357
[c] ACI Committee 202
[d] ACI 318R-89, ACI Building Code

of the weight of cement for plain concrete, 0.04% of the weight of cement for reinforced concrete, and 0.02% of the weight of cement for pre-stressed concrete.

The EU specifications prohibit the use of any additives that have chloride or calcium chloride in reinforced concrete or pre-stressed concrete.

REFERENCES

1988. Demolition and reuse of concrete and masonry reuse of demolition waste. In *Proceedings of the 2nd international symposium RILEM, Building Research Institute and Nihon University, Tokyo*, ed. Y. Kasai, 774. New York: Chapman & Hall.

ACI 228.1R89, In-place methods for determination of strength of concrete. 1994. In *ACI manual of concrete practice, part 2: construction practices and inspection pavements*, 25. Detroit, MI.

Ajdukiewicz, A. B., and A. T. Kliszczewicz. 1999. Utilization of recycled aggregates in HS/HPC. In *5th international symposium on utilization of high strength/high performance concrete*, vol. 2, 973–80. Sandefjord, Norway.

Ajdukiewicz, A.B., and A. T. Kliszczewicz. 2000. Properties and usability of HPC with recycled aggregates. In *Proceedings of the PCI/FHWA/FIP international symposium on high performance concrete*, 89–98. Orlando, FL: Precast/Prestressed Concrete Institute of Chicago.

Ajdukiewicz, A. B., and A. T. Kliszczewicz. 2002. Behavior of RC beams from recycled aggregate concrete. In *ACI fifth international conference*. Cancun, Mexico.

ASTM C114-88: chemical analysis of hydraulic cement.

ASTM C142-78: test method for clay lumps and friable particles in aggregate.

ASTM C183-88: sampling and amount of testing of hydraulic cement.

ASTM C188-84: density of hydraulic cement.

ASTM C349-82: compressive strength of hydraulic cement mortars.

ASTM C670-84: testing of building materials.

ASTM D512-85: standard test method for chloride ion in water.

ASTM D516-82: standard test method for sulfate ion in water.

ASTM D1888-78: standard test method for particulate and dissolved matter, solids or residue in water.

BS 410-1:1986 (2000). Specification for test sieves of metalwire cloth.

BS 812 Part 103:1985. Sampling and testing of mineral aggregate sands and fillers.

BS 882:1992.

BS EN933-1:1997. Tests for geometrical properties of aggregates, determination of particle size distribution. Sieving method.

Di Niro, G., E. Dolara, and R. Cairns. 1998. The use of recycled aggregate concrete for structural purposes in prefabrication. In *Proceedings of the 13th FIP Congress "challenges for concrete in the next millennium," Amsterdam*, vol. 2, 547–50. Rotterdam-Brookfield: Balkema.

ECP203. 2003. Egyptian code for design and execute concrete structures: part 3 laboratory test for concrete materials.

Egyptian standard specification: 76-1989 tension tests for metal.

Egyptian standard specification: 262-1999 steel reinforcement bars.

Egyptian standard specification: 1109-1971 concrete aggregate from natural resources.

Egyptian standard specification: 1947-1991 method of taking cement sample.

Egyptian standard specification: 2421-1993 natural and mechanical properties for cement. Part 1: define cement setting time.

Egyptian standard specification: 2421-1993 natural and mechanical properties for cement. Part 2: define cement fining by sieve no. 170.

Egyptian standard specification: 2421-1993 natural and mechanical properties for cement. Part 2: define cement fining by using Blaine apparatus.

ISO 6274-1982: sieve analysis of aggregate.

Mukai, A. and M. Kikuchi. 1988. Properties of reinforced concrete beams containing recycled aggregate. *In Proceedings of the 2nd international symposium RILEM, Building Research Institute and Nihon University, Tokyo, vol. 2, 670–79.* New York: Chapman & Hall.

Murphy, W. E. 1977. 1977. Contract strength requirements—core versus in situ evaluation. (Discussion on paper by V. M. Malhotra.) *Journal of the American Concrete Institute* 74(10):523–25.

Neville, A. M. 1983. *Properties of concrete.* Pitman.

Plowman, J. M., W. F. Smith, and T. Sheriff. 1974. Cores, cubes, and the specified strength of concrete. *The Structural Engineer* 52(11):421–26.

Salem, R. M., and E. G. Burdette. 1998. Role of chemical and mineral admixtures on physical properties and frost resistance of recycled aggregate concrete. *ACI Materials Journal* 95(5):558–63.

Yuan, R. L., M. Ragab, R. E. Hill, and J. E. Cook. 1991. Evaluation of core strength in high-strength concrete. *Concrete International* 13(5):30–34.

Van Acker, A. 1997. Recycling of concrete at a precast concrete plant. *FIP Notes* Part 1(No. 2):3–6; Part 2(No. 4):4–6.

5 Concrete Mix Design

5.1 INTRODUCTION

Generally, every site location has special characteristics, depending on the aggregate materials that are available near the site. The main task for designing the concrete mix is to obtain the required concrete characteristic strength after 28 days for standard cubes or cylinders according to the project specifications. This is very important in designing the concrete mix; the second factor is the workability and the way of pouring must be known. When using a pump in the casting process a special concrete design mix is needed.

On site, the quality control (QC) team has to accept the concrete mix on site; sometimes a third party can be hired to perform the quality control of the concrete. In any case, the quality control team should be well trained so that by visual inspection only they can determine the quality. In addition, they should have the capability to perform a fast fresh concrete test and compare the results with the project specifications, so they are fully responsible for accepting or refusing the concrete that is delivered to the site from the mixer plant or mixed on site.

The quality control team should understand the basics of concrete design mix in different specifications and have the capability to analyze the data easily by using essential statistical information, as will be clearly explained in this chapter, and compare data with the results and with the project specifications.

5.2 ESSENTIAL STATISTICAL INFORMATION

Some statistics knowledge is essential to the engineer who is responsible for quality in general and specific for concrete quality control, that depends on studying the results of the concrete compressive strength tests. From these results one can define statistically the acceptance and refusal limits. Concrete design mixes in Egyptian, American, and British codes depend on the statistical information.

The essential statistics criteria will be illustrated in the next section, such as the arithmetic mean, standard deviation, and coefficient of variation.

5.2.1 ARITHMETIC MEAN

The arithmetic mean is the average of a group of results and is represented by this equation:

$$\overline{X} = \frac{X_{1+}X_{2+............}+X_n}{n}$$

(5.1)

where

n is the number of the results.

X is the reading for each result.

As a practical example, assume that three samples were taken of standard cubes from the first mixing and a compressive strength test was performed after 28 days. The following results were obtained: 31 N/mm², 30 N/mm², and 29 N/mm². When calculating the arithmetic mean value from the above equation the mean is 30 N/mm².

From the second mixing, three samples were taken of standard cubes and were crushed, also after 28 days under the same conditions as for the previous mixing, and the following results were obtained: 37 N/mm², 30 N/mm², and 23 N/mm². The mean is 30 N/mm².

It is clear that the two mixings provide the same mean value. Does this mean that the two mixings are the same quality? Can we accept the two mixings? From an engineering perspective there are differences, but the arithmetic mean is the same. Therefore, it is important to define another statistics criteria to compare the two mixings.

5.2.2 STANDARD DEVIATION

The standard deviation is the second statistical parameter that expresses the distribution of the test results around the arithmetic mean and is based on the following equation:

$$S = \sqrt{\frac{(X_1 - \overline{X})^2 + (X_1 - \overline{X})^2 + ...+(X_n - \overline{X})^2}{n}}$$

(5.2)

If we apply the above equation to the previous example for the first mixing, the standard deviation will be

$$S = \sqrt{\frac{(31-30)^2 + (30-30)^2 + (29-30)^2}{3}}$$

The standard deviation for the first mixing = 0.816 N/mm². To calculate the standard deviation for the second mixing:

$$S = \sqrt{\frac{(37-30)^2 + (30-30)^2 + (23-30)^2}{3}}$$

The standard deviation for the second mixing = 5.7 N/mm².

The standard deviation of the second mixture has a higher value than the first mixture and the results of the cube's concrete strength in the second mixture are far from the arithmetic mean than the first mixture. This represents a significant deviation from the arithmetic mean, which tells us that quality is very low.

Note that the standard deviation has units. Therefore, it must be conducted through a comparison with the same average value. The previous example presented two mixtures with the same required average concrete compressive strength at 28 days equal to 30 N/mm².

On the other hand, if a comparison of two locations shows different concrete strength, then another factor is required. Assume that the first site required concrete strength 30 N/mm² and the second site has concrete strength equal to 50 N/mm². In that case, the standard deviation has no meaning as a comparison tool, so in this case we need another statistical tool, which is the coefficient of variation.

5.2.3 COEFFICIENT OF VARIATION

The coefficient of variation (C.O.V.) is the true measure of quality control as it determines the proportion of difference or the deviation of the readings from the arithmetic mean (Figure 5.6). This factor has no units, as it is the standard deviation divided by the mean. Therefore, C.O.V. is the main factor for the quality control degree measurement.

$$\text{C.O.V} = \frac{S}{\bar{X}} \qquad\qquad (5.3)$$

As another example, a third mixture was designed at another site to give concrete compressive strength after 28 days values equal to 50 N/mm². The three samples taken gave the compressive strength after 28 days of 51 N/mm², 50 N/mm², and 49 N/mm². When calculating the arithmetic mean and standard deviation, we find that

$$\text{Mean} = 50 \text{ N/mm}^2$$

$$\text{Standard deviation} = 0.816 \text{ N/mm}^2$$

Note that in this site the arithmetic mean is identical with the requirements of the mixture design to have compressive strength after 28 days equal to 50 N/mm². In the previous example, the mean compressive strength was 30 N/mm² with the same value of the standard deviation. Therefore, making the comparison of two sites should be done through calculating the coefficient of variation for each.

$$\text{C.O.V. for first site} = 0.03$$

$$\text{C.O.V. for second site} = 0.02$$

Note that the second site has a lower C.O.V. than the first site, so the percentage of the standard deviation to arithmetic mean is lower in the second location than at

TABLE 5.1

Concrete Strength Values After 28 Days

305	340	298	422	340
267	297	320	382	356
349	366	312	340	355
404	382	306	368	311
350	448	**350**	322	326
303	365	384	346	358
344	339	306	298	398
360	360	282	320	378
352	325	341	326	367
				384

the first site. Therefore, the second site concrete mixture is higher quality that the first site. Whenever C.O.V. is close to zero, quality control is excellent.

5.3   BASICS OF CONCRETE MIX DESIGN

The strength results for 46 cube samples from a particular class of concrete delivered to a project are shown in Table 5.1.

From the descending cumulative curve, one can see that 100% of the sampling results give compressive strength less than 45.9 N/mm² (Figures 5.1–5.3). At the same time, one can see from the results of previous tests that the samples that give results equal to or less than 28 N/mm² are about 2% of the total tested samples (Tables 5.2 and 5.3).

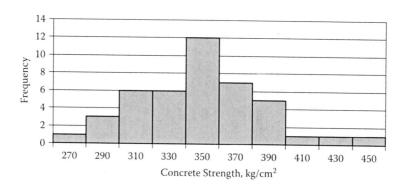

FIGURE 5.1   Frequency histogram.

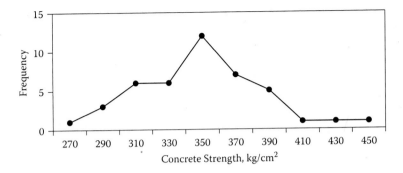

FIGURE 5.2 Relation curve of frequency and concrete strength results.

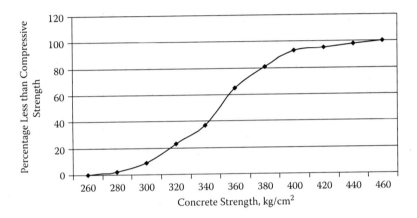

FIGURE 5.3 Descending curve for concrete compressive strength.

TABLE 5.2
Concrete Strength Frequency

Sample No.	Cell Boundaries	Mid-cell Values	Frequency
1	260–280	270	1
2	280–300	290	4
3	300–320	310	6
4	320–340	330	7
5	340–360	350	13
6	360–380	370	7
7	380–400	390	5
8	400–420	410	1
9	420–440	430	1
10	440–460	450	1
		Total	46

Table 5.3
Descending Frequency

Sample No.	Compressive Strength (Kg/cm²)	No. of Readings With Values Lower than the Compressive Strength	Percentage of Readings Less than Lower Level Value
10	460	46	100
10	440	45	98
9	420	41	90
8	400	35	76
7	380	28	61
6	360	15	33
5	340	8	17
4	320	3	6
3	300	2	4
2	280	1	2
1	260	0	0

5.3.1 NORMAL DISTRIBUTION

Normal distribution is the most popoular probability distribution curve as it presents well the most natural phenomena. It is found from the experimental test results indicating that the compressive strength of a concrete cube as a standard test follows the normal distribution. The shape of the normal distribution is shown in Figure 5.4 and it has special characteristics, as follows:

- Normal distribution is symmetrically distributed about the arithmetic mean; on the other hand, the arithmetic mean divides the curve into two equal parts.
- The arithmetic mean, median, and mode are the most likely to coincide.
- The area under the curve is unity since the sum for all probabilities must equal unity.

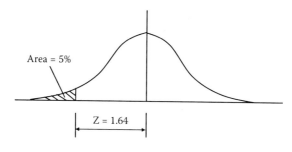

Area = 5%

Z = 1.64

FIGURE 5.4 Normal distribution curve.

- The variable representing cube crushing results can take the values from ∞ to $-\infty$ so the curve presents all the probability values for the concrete compressive strength.

The equation for the probability distribution is as follows:

$$f(x) = \frac{1}{\sigma\sqrt{2\pi}} e^{\frac{(x-\bar{x})^2}{2\sigma^2}} \qquad (5.4)$$

where
  σ is the standrd deviation.
  x is the mean.

Therefore, as a result every distribution shape depends on the mean and the standard deviation, and any variation in these two parameeters affects the shape of the probability distribution. So the standard normal distribution is used to define the area under the curve by knowing the standard deviation and the mean. From the following equation can be obtained another parameter, which is Z:

$$z = \frac{x - \bar{x}}{\sigma} \qquad (5.5)$$

Table 5.4 illustrates the values of the area under the curve by knowing Z as the values of Z in the first column. The first row defines the accuracy to two digital numbers. From this table, one can see that the area under the curve at Z equal to 1.64 is 0.4495. So the area for values higher than Z is 0.5–0.4495, which is equal to 0.0505, which is around 5%. This means that the probability that the result values are less than Z is 5% and that will explain the basics of concrete mix design.

Figure 5.5 shows that the area from the mean to the first value of standard deviation is equal to 34.13% and the area for the two standard deviations is equal to 47.72%.

5.4   EGYPTIAN CODE

The characteristic strength of concrete is defined as the strength below which not more than a prescribed perecentage of the test results should fall. The Egyptian code adopts a percentage of 5%. Hence, by knowing the required concrete characteristic strength f_{cu}, we can define the target strength (f_m) to design the concrete mix, as in the following equation:

$$f_m = f_{cu} + M \qquad (5.6)$$

After designing the concrete mix based on the target strength, the probability of failing the results of the cube strength under the values of the characteristic strength must be less than 5%.

TABLE 5.4
Standard Normal Distribution

Z	0	0.01	0.02	0.03	0.04	0.05	0.06	0.07	0.08	0.09
0	0	0.004	0.008	0.012	0.016	0.0199	0.0239	0.0279	0.0319	0.0359
0.1	0.0398	0.0438	0.0478	0.0517	0.0557	0.0596	0.0636	0.0675	0.0714	0.0753
0.2	0.0793	0.0832	0.0871	0.091	0.0948	0.0987	0.1026	0.1064	0.1103	0.1141
0.3	0.1179	0.1217	0.1255	0.1293	0.1331	0.1368	0.1406	0.1443	0.148	0.1517
0.4	0.1554	0.1591	0.1628	0.1664	0.17	0.1736	0.1772	0.1808	0.1844	0.1879
0.5	0.1915	0.195	0.1985	0.2019	0.2054	0.2088	0.2123	0.2157	0.219	0.2224
0.6	0.2257	0.2291	0.2324	0.2357	0.2389	0.2422	0.2454	0.2486	0.2517	0.2549
0.7	0.258	0.2611	0.2642	0.2673	0.2704	0.2734	0.2764	0.2794	0.2823	0.2852
0.8	0.2881	0.291	0.2939	0.2967	0.2995	0.3023	0.3051	0.3078	0.3106	0.3133
0.9	0.3159	0.3186	0.3212	0.3238	0.3264	0.3289	0.3315	0.334	0.3365	0.3389
1	0.3413	0.3438	0.3461	0.3485	0.3508	0.3531	0.3554	0.3577	0.3599	0.3621
1.1	0.3643	0.3665	0.3686	0.3708	0.3729	0.3749	0.377	0.379	0.381	0.383
1.2	0.3849	0.3869	0.3888	0.3907	0.3925	0.3944	0.3962	0.398	0.3997	0.4015
1.3	0.4032	0.4049	0.4066	0.4082	0.4099	0.4115	0.4131	0.4147	0.4162	0.4177
1.4	0.4192	0.4207	0.4222	0.4236	0.4251	0.4265	0.4279	0.4292	0.4306	0.4319
1.5	0.4332	0.4345	0.4357	0.437	0.4382	0.4394	0.4406	0.4418	0.4429	0.4441
1.6	0.4452	0.4463	0.4474	0.4484	<u>0.4495</u>	0.4505	0.4515	0.4525	0.4535	0.4545
1.7	0.4554	0.4564	0.4573	0.4582	0.4591	0.4599	0.4608	0.4616	0.4625	0.4633
1.8	0.4641	0.4649	0.4656	0.4664	0.4671	0.4678	0.4686	0.4693	0.4699	0.4706
1.9	0.4713	0.4719	0.4726	0.4732	0.4738	0.4744	0.475	0.4756	0.4761	0.4767
2	0.4772	0.4778	0.4783	0.4788	0.4793	0.4798	0.4803	0.4808	0.4812	0.4817
2.1	0.4821	0.4826	0.483	0.4834	0.4838	0.4842	0.4846	0.485	0.4854	0.4857
2.2	0.4861	0.4864	0.4868	0.4871	0.4875	0.4878	0.4881	0.4884	0.4887	0.489
2.3	0.4893	0.4896	0.4898	0.4901	0.4904	0.4906	0.4909	0.4911	0.4913	0.4916
2.4	0.4918	0.492	0.4922	0.4925	0.4927	0.4929	0.4931	0.4932	0.4934	0.4936
2.5	0.4938	0.494	0.4941	0.4943	0.4945	0.4946	0.4948	0.4949	0.4951	0.4952
2.6	0.4953	0.4955	0.4956	0.4957	0.4959	0.496	0.4961	0.4962	0.4963	0.4964
2.7	0.4965	0.4966	0.4967	0.4968	0.4969	0.497	0.4971	0.4972	0.4973	0.4974
2.8	0.4974	0.4975	0.4976	0.4977	0.4977	0.4978	0.4979	0.4979	0.498	0.4981
2.9	0.4981	0.4982	0.4982	0.4983	0.4984	0.4984	0.4985	0.4985	0.4986	0.4986
3	0.4987	0.4987	0.4987	0.4988	0.4988	0.4989	0.4989	0.4989	0.499	0.499

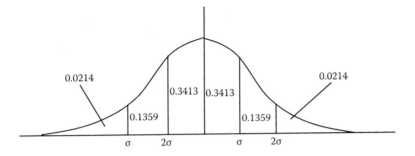

FIGURE 5.5   Dividing area under normal distribution.

TABLE 5.5
Safety Margin Factor for Concrete Design Mix

Availability of Test Results	Margin Safety Factor, M, for Concrete Compressive Strength, f_{cu}		
	$f_{cu} < 200$	$200 \leq f_{cu} < 400$	$600 \geq f_{cu} \geq 400$
1. Available 40 results or more with similar materials and condition	(1.64 S) and not less than 4 N/mm²	(1.64 S) and not less than 6 N/mm²	(1.64 S) and not less than 7.5 N/mm²
2. No available data or fewer than 40 test[a] results with similar materials and condition	Not less than 0.6 f_{cu}	Not less than 12 N/mm²	Not less than 15 N/mm²

[a]   The test presents average of 3 standard cubes taken from same mix.

M is the safety factor to verify that the perecentage of the crushed cube's strength values less than f_{cu} will not be less than 5% (Table 5.5). This safety factor is a function of the standard deviation, as shown in the following equation:

$$f_m = f_{cu} + 1.64\ S \qquad (5.7)$$

Table 5.6 provides a guideline to predict the standard deviation of the concrete after knowing the QC of the site by visiting the site only; by performing a test and calculating the standard deviation one can audit and categorize the work and supervising activities of the QC in this site.

This is also is a guide for the QC indicator for the ready mix batch plant that supplies the concrete to your site.

Table 5.7 is from ACI 214-77 and presents the overall standard deviation for concrete in laboratory trial mixes and in the field for concrete strength 35 MPa.

TABLE 5.6
Expected Standard Deviation Values

Quality Control Condition	Standard Deviation (N/mm²)
Good QC with continuous supervision	4–5
Moderate QC with occasional supervision	5–7
Poor QC with no supervision	7–9

TABLE 5.7
Classification of Standards of Concrete Based on ACI 214-77

Standard of Control	Overall Standard Deviation (MPa)	
	In Field	Laboratory Trial Mixes
Excellent	<3	<1.5
Very good	3–3.5	1.5
Good	3.5–4	1.5–2
Fair	4–5	2–2.5
Poor	>5	>2.5

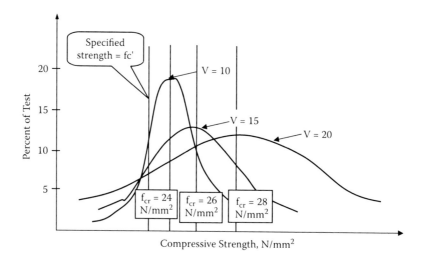

FIGURE 5.6   Normal frequencies curve for C.O.V. 10, 15, 20.

5.5   BRITISH STANDARD

The ECP has similar requirements to BS 5328:part 4:1990. The British practice cubes are also used. The British approach is to use a charactristic strength, defined as the value of strength below which 5% of all possible test results are expected to fall; the margin between the charactristic strength and the mean strength is selected to verify this probability. The following criteria must be applied to achieve this probability:

1. The average value of any four consecutive test results exceeds the specified charactristic strength by 3 MPa.
2. No test results fall below the specified characteristic strength by more than 3 MPa.

Similar requirements are prescribed for the flexural test: the values in criteria 1 and 2 are then 0.3 MPa.

5.6   AMERICAN SPECIFICATION (ACI)

ACI code states that the concrete production facility has a record of at least 30 consecutive strength tests representing materials and conditions similar to those expected; the strength used as the basis for selecting concrete proportions must be the larger of

$$f_{cr} = f_c + 1.34 \, S \tag{5.8}$$

or

$$f_{cr} = f_c + 2.33 \, S - 500 \tag{5.9}$$

where
f_{cr} is the required target strength in preparing the concrete mix.
f_c is the concrete strength after 28 days.

If the standard deviation is unknown, the required average strength f_{cr} used as the basis for selecting concrete proportions must be determined from the following:

$$f_{cr} = f_c + 7 \text{ N/mm}^2 \, f_c < 21 \text{ N/mm}^2$$

$$(f_{cr} = f_c + 1000 \text{ psi } f_c < 3000 \text{ psi})$$

$$f_{cr} = f_c + 8.4 \text{ N/mm}^2 \, f_c = 21 - 35 \text{ N/mm}^2$$

$$(f_{cr} = f_c + 1200 \text{ psi } f_c = 3000 - 5000 \text{ psi })$$

$$f_{cr} = f_c + 9.8 \text{ N/mm}^2 \, f_c > 35 \text{ N/mm}^2$$

$$(f_{cr} = f_c + 1400 \text{ psi } f_c > 5000 \text{ psi})$$

Formulas for calculating the required target strengths are based on the following criteria:

1. A probability of 1% that an average of three consecutive strength tests will be below the specified strength, f_c ($f_{cr} = f_c + 1.34 S$)
2. A probability of 1% that an individual strength test will be more than 3.5 N/mm^2 (500 psi) below the specified strength fc($f_{cr} = f_c + 2.33 S - 500$)

Criterion 1 will produce a higher required target strength than criterion 2 for low to moderate standard deviations, up to 500 psi. For higher standard deviations criterion 2 will govern.

5.6.1   ACCEPTANCE AND REFUSAL FOR CONCRETE MIX

After a mix is approved for the proposed project, the concrete received on site will be acceptable if the results of the crushing standard cylinder tests after 28 days meet both of the following:

1. No single test strength or the average strengths of two cylinders from a batch shall be more than 3.5 N/mm^2 (500 psi) below the specified compressive strength f_c, i.e., 21.1 N/mm^2 (3000 psi) for specified 24.6 N/mm^2 (3500 psi) concrete.
2. The average of any three consecutive test strengths must equal or exceed the specified compressive strength, f_c.

5.6.2   CONCRETE MIX PROCEDURE

In summary, the mixing procedure is stated in ACI 211.1. Estimating the required batch weights for the concrete involves a sequence of logical, straightforward steps, which in effect fit the characteristics of the available materials into a mixture suitable for the work. The question of suitability is frequently not left to the individual selecting the proportions. The job specifications may dictate some or all of the following:

- Maximum water-cement or water-cementitious material ratio
- Minimum cement content
- Air content
- Slump
- Maximum size of aggregate
- Strength
- Other requirements relating to strength over design, admixtures, and special types of cement, other cementitious materials, or aggregate

Regardless of whether the concrete characteristics are prescribed by the specifications or left to the individual selecting the proportions, establishment of batch weights per cubic meter of concrete can be best accomplished in the following sequence:

TABLE 5.8
Recommended Slumps for Various Types of Construction

Types of Construction	Slump (mm)	
	Maximum[a]	Minimum
Reinforced foundation, walls, and footings	75	25
Plain footings, caissons, and substructure walls	75	25
Beams and reinforced walls	100	25
Building columns	100	25
Pavement and slabs	75	25
Mass concrete	50	25

[a] Slump may be increased when chemical admixtures are used, provided that the admixture-treated concrete has the same or lower w/c ratio and does not exhibit segregation potential or excessive bleeding.

5.6.2.1 Step 1: Choice of Slump

If slump is not specified, a value appropriate for the work can be selected from Table 5.8. The slump ranges shown apply when vibration is used to consolidate the concrete. Mixes of the stiffest consistency that can be placed efficiently should be used. Note that the values of the slump listed in Table 5.8 can be increased when chemical admixtures are used.

5.6.2.2 Step 2: Choice of Maximum Size of Aggregate

Large nominal maximum sizes of well-graded aggregates have fewer voids than smaller sizes. Hence, concretes with the larger-sized aggregates require less mortar per unit volume of concrete. Generally, the nominal maximum size of aggregate should be the largest that is economically available and consistent with dimensions of the structure. In no event should the nominal maximum size exceed one fifth of the narrowest dimension between sides of forms, one third the depth of slabs, nor three fourths of the minimum clear spacing between individual reinforcing bars, bundles of bars, or pre-tensioning strands. These limitations are sometimes waived if workability and methods of consolidation are such that the concrete can be placed without honeycomb or void.

In areas congested with reinforcing steel, post-tension ducts, or conduits, select a nominal maximum size of the aggregate so concrete can be placed without excessive segregation, pockets, or voids. When high strength concrete is desired, best results may be obtained with reduced nominal maximum sizes of aggregate since these produce higher strengths at a given water/cement (w/c) ratio.

5.6.2.3 Step 3: Estimation of Mixing Water and Air Content

The quantity of water per unit volume of concrete required to produce a given slump is dependent on the nominal maximum size, particle shape, and grading of the

TABLE 5.9

Approximate Mixing Water (kg/m³) for Different Slumps and Nominal Maximum Sizes of Aggregate for Non–Air Entrained Concrete Based on ACI

Slump (mm)	Nominal Maximum Aggregate size (mm)							
	10	12.5	19	25	37.5	50	75	152
25–50	206	198	186	177	162	153	130	112
75–100	227	215	201	192	177	168	145	124
152–178	242	227	212	201	185	177	159	—

Approximate Amount of Entrapped Air in Non–Air
Entrained Concrete (%)

3	2.5	2	1.5	1	0.5	0.3	0.2

[a] Rounded aggregate will generally require 13.5 kg less water for non–air entrained.

[b] The use of water reducing chemical admixture, ASTM C494, may also reduce mixing water by 5% or more.

[c] The slump values of more than 178 mm are only obtained through the use of water-reducing chemical admixtures; they are for concrete containing nominal maximum size aggregate not larger than 25 mm.

aggregates; the concrete temperature; the amount of entrained air; and use of chemical admixtures.

Slump is not greatly affected by the quantity of cement or cementitious materials within normal use levels (under favorable circumstances the use of some finely divided mineral admixtures may lower water requirements slightly).

Table 5.9 provides estimates of required mixing water for concrete made with various maximum sizes of aggregate, with and without air entrainment. Depending on aggregate texture and shape, mixing water requirements may be somewhat above or below the tabulated values, but they are sufficiently accurate for the first estimate. The differences in water demand are not necessarily reflected in strength since other compensating factors may be involved.

Rounded and angular coarse aggregates, both of which are similarly graded and of good quality, can be expected to produce concrete of about the same compressive strength for the same cement factor in spite of differences in w/c ratio resulting from the different mixing water requirements.

Particle shape is not necessarily an indicator that an aggregate will be either above or below its strength-producing capacity.

5.6.2.4   Step 4: Chemical Admixtures

Chemical admixtures modify the properties of concrete to make it more workable, durable, and/or economical; increase or decrease the time of set; accelerate strength gain; and/or control temperature gain.

Chemical admixtures should be used only after an appropriate evaluation has been conducted to show that the desired effects have been accomplished under the conditions of intended use. Making sure that water reducing and/or set-controlling admixtures conform to the requirements of ASTM C494, when used singularly or in combination with other chemical admixtures, will significantly reduce the quantity of water per unit volume of concrete.

The use of some chemical admixtures, even at the same slump, will improve such qualities as workability, finish ability, pump-ability, durability, and compressive and flexural strength. Significant volume of liquid admixtures should be considered as part of the mixing water. The slumps shown in Table 5.8 from ACI may be increased when chemical admixtures are used, providing the admixture-treated concrete has the same or a lower water/cement ratio and does not exhibit segregation potential and excessive bleeding. When only used to increase slump, chemical admixtures may not improve any of the properties of the concrete.

Table 5.9 indicates the approximate amount of entrapped air to be expected in non–air-entrained concrete and the recommended average air content for air-entrained concrete. If air entrainment is desired, three levels of air content are given for each aggregate size, depending on the purpose of the entrained air and the severity of exposure if entrained air is needed for durability.

5.6.2.4.1 Mild Exposure

When air entrainment is desired for a beneficial effect other than durability, such as to improve both the workability and cohesion or in low-cement-factor concrete to improve strength, air contents lower than those needed for durability can be used. This exposure includes indoor or outdoor service in a climate where concrete will not be exposed to freezing or to de-icing agents.

5.6.2.4.2 Moderate Exposure

This refers to a climate where freezing is expected but where the concrete will not be continually exposed to moisture or free water for long periods prior to freezing and will not be exposed to de-icing agents or other aggressive chemicals. Examples include exterior beams, columns, walls, girders, or slabs that are not in contact with wet soil and are so located that they will not receive direct applications of de-icing salts.

5.6.2.4.3 Severe Exposure

Concrete that is exposed to de-icing chemicals or other aggressive agents or where the concrete may become highly saturated by continued contact with moisture or free water prior to freezing. For example, pavements, bridge decks, curbs, gutters, sidewalks, canal linings, or exterior water tanks or sumps.

The use of normal amounts of air entrainment in concrete with a specified strength around 35 N/mm² (5000 psi) may not be possible due to the fact that each added percent of air lowers the maximum strength obtainable with a given combination of materials. In these cases the exposure to water, de-icing salts, and freezing temperatures should be carefully evaluated. If a member is not continually wet and will not be exposed to de-icing salts, lower air-content values such as those given in

Table 5.9 for moderate exposure are appropriate even though the concrete is exposed to freezing and thawing temperatures.

However, for an exposure condition where the member may be saturated prior to freezing, the use of air entrainment should not be sacrificed for strength. In certain applications, it may be found that the content of entrained air is lower than that specified, despite the use of usually satisfactory levels of air-entraining admixture.

This happens occasionally, for example, when very high cement contents are involved. In such cases, the achievement of required durability may be demonstrated by satisfactory results of examination of air-void structure in the paste of the hardened concrete.

When trial batches are used to establish strength relationships or verify strength-producing capability of a mixture, the least favorable combination of mixing water and air content should be used. The air content should be the maximum permitted or likely to occur, and the concrete should be gagged to the highest permissible slump. This will avoid developing an overly optimistic estimate of strength on the assumption that average rather than extreme conditions will prevail in the field. If the concrete obtained in the field has a lower slump and/or air content, the proportions of ingredients should be adjusted to maintain required yield. For additional information on air content recommendations, see ACI 201.2R, 301, and 302.1R.

5.6.2.5   Step 5: Selection of Water/Cement (w/c) Ratio

The required w/c ratio is determined not only by strength requirements but also by factors such as durability. Since different aggregates, cements, and cementitious materials generally produce different strengths in the same amount of water, it is highly desirable to know the relationship between strength and w/c for the materials actually to be used. In the absence of such data, approximate and relatively conservative values for concrete containing Type I Portland cement can be taken from Table 5.10 with typical materials and the tabulated w/c ratio should produce the strengths shown, based on 28-day tests of specimens cured under standard laboratory conditions. The average strength selected must, of course, exceed the specific strength by a sufficient margin to keep the number of low tests within specific limits.

For severe conditions of exposure, the w/c ratio should be kept low even though strength requirements may be met with a higher value. Table 5.11 gives limiting values. When natural pozzolans, fly ash, GGBF slag, and silica fume, hereafter referred to as pozzolanic materials, are used in concrete, a water/cement plus pozzolanic materials ratio (or water/cement plus other cementitious materials ratio) by weight must be considered in place of the traditional water/cement ratio by weight.

5.6.3   Mix Proportions

The selection of the concrete mix depends on the available aggregate near the site. So the properties of these aggregate should be defined. The main factor in any engineering practice is the project economy so the selection of the mix proportion should verify the following:

TABLE 5.10
Relation of w/c Ratio and Concrete Compressive Strength

Compressive Strength at 28 days

(N/mm²)	Non–Air Entrained Concrete	Air Entrained Concrete
42	0.41	—
35	0.48	0.40
28	0.57	0.48
21	0.68	0.59
14	0.82	0.74

Notes: Values are estimated average strengths for concrete of not more than 2% air for non–air entrained concrete and 6% total air content for air entrained concrete. For a constant w/c, the strength of concrete is reduced as the air content is increased. Twenty-eight-day strength values may be conservative and may change when various cementitious materials are used. The rate at which the 28-day strength is developed may also change. Strength is based on 6 × 12 in cylinders moist-cured for 28 days in accordance with the sections on "Initial Curing" and "Curing of Cylinders for Checking the Adequacy of Laboratory Mixture Proportions for Strength or as the Basis for Acceptance or for Quality Control" of ASTM method C31 for Making and Curing Concrete Specimens in the Field. The relationship in this table assumes a nominal maximum aggregate size of about 3/4 to 1 in. For a given source of aggregate, strength produced at a given w/c will increase as nominal maximum size of aggregate decreases.

TABLE 5.11
Maximum Permissible w/c Ratio for Concrete in Severe Exposures[a]

Type of Structure	Structure Wet Continuously or Frequently and Exposed to Freezing and Thawing[b]	Structure Exposed to Sea Water or Sulfates
Thin sections (railings, curbs, sills, ledges, ornamental work) and sections with less than 25 mm cover over steel	0.45	0.40[b]
All other structures	0.50	0.45[b]

[a] Based on report of ACI Committee 201. Cementitious materials other than cement should conform to ASTM C618 and C989.

[b] If sulfate resisting cement (Type II or Type V of ASTM C150) is used, water-cementitious materials ratio may be increased by 0.05.

- The required concrete characteristic compressive strength
- The durability of concrete, which requires defining the w/c ratio and cement content
- The optimization of availability of materials, performance, and economics

The following equation defines the concrete mix:

$$\frac{C}{1000\gamma_c} + \frac{A_f}{1000\gamma_f} + \frac{Ag}{1000\gamma_g} + \frac{W}{1000} = 1 \qquad (5.10)$$

where
    C = mass of cement.
    A_f = mass of fine aggregate.
    A_g = mass of coarse aggregate.
    W = mass of water.
    $W = (w/c) \times C$.
    γ_c = cement specific gravity.
    γ_f = fine aggregate specific gravity.
    γ_g = coarse aggregate specific gravity.

This equation calculates the quantities of ingredients to produce 1 cubic meter of concrete. Note that the cement content proposed is based on the environmental conditions, and the required concrete strength can define the water/cement ratio (w/c). From experience, the required mixing ratio can define the cement/aggregate ratio $[C/(A_f + A_g)]$, and also the coarse/fine aggregate ratio (A_f/A_g).

Hence, from these ratios and the previous equation the mix proportions can be defined.

5.6.3.1 British Standard

The British standard BS8110 provides a guide to suitable concrete mixtures based on the environmental conditions classified as shown in Table 5.12.

After defining the environmental conditions based on BS8110, described in Table 5.12, refer to Table 5.13, which provides the minimum concrete grade required with a minimum cement content based on the maximum aggregate size and the corresponding maximum w/c ratio.

5.7   FRESH CONCRETE TEST

5.7.1   CYLINDER AND CUBE TEST

Three types of compression test specimens are used: cubes, cylinders, and prisms. Cubes are used in Great Britain, Germany, and many other countries in Europe. Cylinders are the standard specimens in the United States, France, Canada, Australia, and New Zealand. In Scandinavia tests are done on both cubes and cylinders.

TABLE 5.12

Classification of Environmental Conditions Based on BS8110 Part 1-1985

Environment	Exposure Conditions
Mild	Concrete surfaces protected against weather or aggressive conditions
	Exposed concrete surfaces but sheltered from severe rain or freezing while wet
	Concrete surfaces continuously under nonaggressive water
	Concrete in contact with nonaggressive soil
Moderate	Concrete subject to condensation
Severe	Concrete surfaces exposed to severe rain, alternate wetting and drying, or occasional freezing or severe condensation
	Concrete surfaces occasionally exposed to sea water spray or de-icing salts (directly or indirectly)
Very severe	Concrete surfaces exposed to corrosive fumes or severe freezing conditions while wet
	Concrete surfaces frequently exposed to sea water spray or de-icing salts (directly or indirectly)
Most severe	Concrete in sea water tidal zone down to 1 m below lowest low water
Abrasive[a]	Concrete surfaces exposed to abrasive action, e.g., machinery, metal tiered vehicles, or water carrying solids

Note: (1) For aggressive soil and water conditions see 5.3.4 of BS 5328-1:1997. (2) For marine conditions see also BS 6349.

[a] For flooring see BS 8204.

The tendency nowadays, especially in research, is to use cylinders in preference to cubes, but before comparing the two types of specimens the various tests should be considered in detail.

5.7.1.1 Cube Test

The specimens are cast in steel or cast-steel molds, generally 150 mm cubes. The standard practice prescribed by BS1881:Part 3:1970 is to fill the mold in three layers. Each layer of concrete is compacted by not less than 35 strokes of a 25 mm (1 in) square steel rod. Ramming should continue until sufficient compaction has been achieved, for it is essential that the concrete in the cube be fully compacted if the compressive test is to be representative of the properties of fully compacted concrete.

After the top surface of the cube has been finished by trowel, the cube is stored undisturbed for 24 hours at 18°C to 22°C and relative humidity of not less than 90%. At the end of this period the mold is stripped and the cube is further cured in water at 19°C to 21°C.

The test is generally performed at 28 days but additional tests are also performed at 3 and 7 days. In the compression test, the cube is placed with the cast faces in contact with the platens of the testing machine, i.e., the position of the cube when tested is at right angles to that as-cast. It is worth mentioning that according to BS1881:Part 4:1970 the load on the cube should be applied at a constant rate of stress equal to 15 MPa/min (2200 psi/min).

TABLE 5.13

Requirements of BS 8110: Part 1-1985 to Ensure Durability under Specific Conditions of Exposure of Plain Concrete

Exposure Conditions	Max. w/c	Min. Grade MPa	Min. Content of Cement for Max. Nominal Aggregate Size (kg/m³)			
			40 mm	20 mm	14 mm	10 mm
Mild	0.80	20	150	180	200	220
Moderate	0.65	30	245	275	295	315
Severe	0.60	35	270	300	320	340
Very severe	0.55	35	295	325	345	365
Extreme	0.50	45	320	350	370	390

The cube standard 150 mm is more common and there are some other sizes as in the Table 5.13.

5.7.1.2 Cylinder Test

The standard cylinder size is 150 mm diameter by 300 mm height or 100 mm diameter by 200 mm height and it is cast in a mold generally made of steel or cast steel. Use a cylinder 100 mm diameter in the case of maximum nominal aggregate size, and not higher than 20 mm and 40 mm in the case of 150 mm diameter. The cylinder specimens are made in a similar way to the cubes but are compacted either in three layers using a 16 mm diameter rod or in two layers by means of an immersion vibrator. Details of the procedure are described in ASTM Standard C192-76. The preparation of the cylinder is as shown in Figure 5.7.

The top surface of the cylinder finished with a float is not smooth enough for testing and requires further preparation: this is the greatest disadvantage of this type of specimen as normally used.

The top surface can be prepared in two ways. The first method is by using mortar on the top by using a collar with handle and filling by cement mortar, and the second method prepares the surface top cover by adding sulfur and fine sand with small amount of carbon (1%–2%) and this composition is heated to 130°C–150°C and cools slightly.

FIGURE 5.7   Cylinder preparation.

The cylinder is placed on a layer of coverage as thin as possible to make sure that the vertical axis of the cylinder, and then cut the excess after several seconds and center of the surface of the cylinder. During the test you must make sure that the cylinder cover will not slide or break before the collapse of the sample.

This test relies on sampling of concrete during the pouring process and must be represented as much as possible by a sample of concrete cast into structural members.

The samples will be taken by a standard shovel manufactured from non-rust material with thickness 0.8 mm. The amount of concrete taken by shovel at a time is about 5 kg and the number of shovels is determined in accordance with the standard type of test. Take into account when sampling from the mixer on site or from the mixer truck of ready mix concrete to exclude the first and last part of the shipment and never take samples that are not good representatives of the whole batch.

In the case of putting concrete in a horizontal position before delivering it for casting must sampling of places allocated to the surface and with depth as you can.

In the case of casting, samples must be taken out of the mixer or from the mixing truck using cranes or pumps.

5.7.2 Predict Concrete Strength

The number of samples taken is according to the prediction of the concrete strength and it is often done at 7 days and 28 days. As shown in Table 5.14, according to the Egyptian code, predicting the concrete strength at 28 days can be done by knowing the cube strength at 7 and 3 days. The number of cubes or cylinders tested at 3 and 7 days should be stated in the project specifications, which can also specify the minimum strength accepted at 3 or 7 days and define the reasonable time to remove the wood or the steel forms.

In the past, the gain in strength beyond the age of 28 days was regarded merely as contributing to an increase in the factor of safety of the structure, but since 1957 the code of practice for reinforced and prestressed concrete allows the gain in strength to be taken into account in the design of structures that will not be subjected to load until a later age except when no-fines concrete is used; with some lightweight aggregates, verifying tests are advisable. The values of strength given in the British Code of Practice CP110:1972, based on the 28-day compressive strength, are given in Table 5.15. Note that this table will not apply when accelerators are used.

TABLE 5.14
Correction Factor for Concrete Compressive Strength Test Results

	Concrete Age (Days)				
Type of Cement	3	7	28	90	360
Ordinary Portland cement	2.5	1.5	1.0	0.85	0.75
Fast hardening Portland cement	1.8	1.2	1.0	0.9	0.85

TABLE 5.15

British Code of Practice CP 110:1972 Factors for Increase in Compressive Strength of Concrete with Age (Average Values)

Minimum Age of Member When Full Design Load Is Applied (Months)	Age Factors for Concrete with 28-Day Strength (MPa)		
1	1.00	1.00	1.00
2	1.10	1.09	1.07
3	1.16	1.12	1.09
6	1.20	1.17	1.13
12	1.24	1.23	1.17

TABLE 5.16

Correction Factors for Concrete Strengths for Different Molds

Cube	Mold Dimensions (mm)	Correction Factor
Cube	100 × 100 × 100	0.97
	150 × 150 × 150	1.00
Cube	158 × 158 × 158	
Cube	200 × 200 × 200	1.05
Cube	300 × 300 × 300	1.12
Cylinder	100 × 200	1.20
Cylinder	150 × 300	1.25
Cylinder	250 × 500	1.30
	150 × 150 × 300	1.25
Prism	158 × 158 × 316	
	150 × 150 × 450	1.30
Prism	158 × 158 × 474	
Prism	150 × 150 × 600	1.32

As mentioned before, the British standard and some countries use cubes, and in America and other countries the specifications say to use cylinders. Table 5.16 defines the relation between cube and cylinder test values so if you have the values of cylinder compressive strength it can be converted to cube value and vice versa.

5.8   DEFINE CONCRETE DENSITY

Concrete density is an important factor for determining the dead loads for an accurate structure analysis (Figures 5.8 and 5.9). By knowing the concrete density one can predict the concrete compressive strength and the permeability.

FIGURE 5.8   Compressive strength and density machines test.

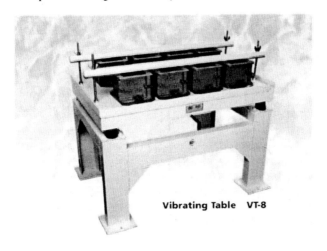

FIGURE 5.9   Vibration table.

In some special structures the main factor in the structure is the concrete density and for a heavy concrete structure affected by floating force, producing dense concrete is very critical. The relation between the density and compressive strength is shown in Figure 5.9.

Samples should be chosen so that the size is not less than 50 S^3, where S is nominal maximum aggregate size; sample size should not be less than 0.001 m³.

Often the sample is in the form of a cube or cylinder, for which volume (V) is accurately determined by the nearest millimeter dimensions.

The sample weight (w_1) is taken upon arrival at the laboratory.

Measure the weight of the sample after immersion in water at a temperature of 20°C + 2°C until the two consequent reading weights are approximately the same, and the time difference between 24 hours and the weight is appropriate if the change does not exceed 0.02% and the weight is W_2.

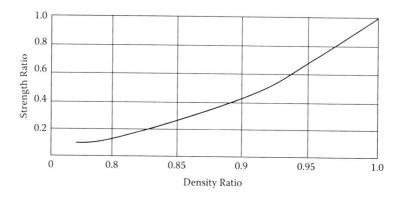

FIGURE 5.10   Relation between density and compressive strength. *Source:* Adopted from Neville, A. M. 1983. *Properties of concrete.* Pitman.

Measure the weight of the sample (W_d) at dry condition by drying the sample in an oven at a temperature of 105°C + 5°C until two consequent reading weights are approximately the same, and the time difference between 24 hours and the weight is appropriate if the change does not exceed 0.02%.

Calculate density in the three cases:

$$D_1 = W_1/V \text{ (laboratory sample)}$$

$$D_2 = W_2/V \text{ (immersing sample)}$$

$$D_3 = W_d/V \text{ (dry sample)}$$

Based on ACI-211.1-91 estimate the density of fresh concrete for different maximum aggregate size in the case of air entrained and non-air entrained concretes (Table 5.17).

TABLE 5.17

First Estimation of Fresh Concrete Density

Maximum Size of Aggregate (mm)	Non-Air Entrained (Kg/m³)	Air Entrained
10	2285	2190
12.5	2315	2235
20	2355	2280
25	2375	2315
40	2420	2355
50	2445	2375
70	2465	2400
150	2505	2435

FIGURE 5.11   Slump test tools.

5.9   DEFINE SETTLEMENT FOR FRESH CONCRETE

The slump test is easy to carry out on concrete and gives good information on the concrete before casting; therefore, it is considered one of the key means to control the quality of concrete at the site. It is therefore widely used.

The slump test is prescribed by ASTM C143-90a and BS1881:Part 102:1983. The tool used in this test is a metal template cone with a steel flat plate and a steel rod for compaction. The materials these tools are made from must not be affected by the cement paste. Their thickness should be at least 1.5 mm and the entire surface smooth and free from any nails or juts. The tools are shown in Figure 5.11.

All the tools have standard dimensions such as the cone dimensions (Figure 5.12), as follows:

Top diameter = 100 ± 2 mm.
Bottom diameter = 200 ± 2 mm.
Height = 300 ± 2 mm.

The flat basin for preparing the sample has dimensions 1.2 × 1.2 m and depth equal to 50 mm and thickness equal to 1.6 mm. The steel rod for compaction is round with 16 mm diameter and length of 600 mm and endings semi-spherical.

This test is performed after mixing the concrete and when pouring. Three cubic meters are taken by six shovels that are clean and of standard size. Finish a sample in the flat basin, and mix twice using the shovels. Then prepare the horizontal surface so it is clean and flat to hold the cone.

Pour concrete inside the cone to one third of the cone's height and compact it using the standard rod 25 times until it is distributed equally inside the concrete mold (Figure 5.13). Take into account that only the first floor bar compaction contacts with the horizontal surface. Then pour the second layer and perform

FIGURE 5.12   Slump cone dimensions.

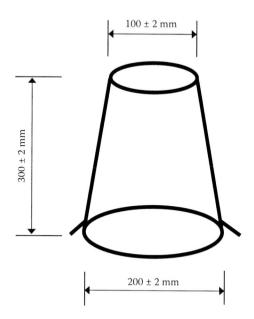

FIGURE 5.13   Pouring concrete in slump.

FIGURE 5.14 Measuring the slump settlement.

compaction by the same way. Finally, pour the last third and perform compaction using the same precautions and the last surface will be finished by trolling.

Remove the mold vertically, slowly, and carefully after 5 to 10 seconds. Note that the entire process from the start of placing the concrete until full lifting of the template should require not more than about 150 seconds. If the collapse of the sample happens, repeat the test again.

Put the steel rod on the top horizontal to the converted cone and measure the settlement of the concrete, as shown in Figure 5.14, and compare to the project specifications. According to specifications set by the Egyptian code, allowable settle and allowable tolerance in the maximum allowable slump settlement must be stated in project specifications or defined as shown in Tables 5.18, 5.19 and 5.20.

The slump test is very useful on site for monitoring the concrete quality day by day and hour by hour due to variation in the materials fed into the mixer. Too much or too little slump gives immediate warning and enables the mixer operator to remedy the situation.

TABLE 5.18
Required Slump Values for Different Concrete Members

Element Type	Slump (mm)	Type of Compaction
Concrete blocks	0–25	Mechanical
		Mechanical
Concrete foundation with light reinforcement and medium reinforcement and concrete section with light reinforcement	25–50	Mechanical
Concrete section with medium or high reinforcement	50–100	Manual
Concrete section with dense reinforcement	100–125	Light
Deep foundation and pumped concrete	125–200	Light

TABLE 5.19
Allowable Tolerance in Defining Maximum Slump Settlement

Maximum Amount of Settlement (mm)	Allowable Tolerance (mm)
75 or less	35
>75	60

TABLE 5.20
Allowable Tolerance in Defining Required Slump Settlement

Required Settlement Value (mm)	Allowable Tolerance (mm)
50 or less	±10
50–100	±20
>100	±30

5.10   DETERMINING COMPACTING FACTOR FOR FRESH CONCRETE

The compacting factor test is described in BS1881:Part 103:1993 and in ACI 211.3-75 (revised 1987 and reapproved 1993).

This test will be done to determine a working compaction of concrete interoperability with a low or medium weight and will be applied to regular concrete manufactured with air entrained for aggregate with normal, light, or heavy weight. Therefore, any nominal maximum aggregate size is not larger than 40 mm and often this test is only performed in precast concrete or at large work sites.

The apparatus is shown in Figure 5.15. It consists of two cones above a cylinder fixed on a steel support. The dimensions of the cylinder, cone, and the distance between them are based on BS1881.

Gently place a sample of concrete in the upper cone to the level of the edge, then open the gate of the upper cone to let the concrete fall on the lower cone. Then open the gate of the lower cone to allow the concrete to fall down over the cylinder.

The ruler is measured in part by the weight of partially compacted concrete (W_1) to the nearest 10 grams within 150 seconds of the start time of the test.

Then refill the cylinder with the same type of concrete to be fully compacted and weigh the concrete to be (W_2) to the nearest 10 g. The workability of the concrete will be shown in Table 5.21.

$$\text{Compaction factor} = W_1/W_2$$

FIGURE 5.15   Compaction factor apparatus.

TABLE 5.21
Compaction Factor and Workability

Concrete Workability Degree	Compaction Factor
Very low	0.8
Low	0.87
Moderate	0.935
High	0.96

5.11   HIGH PERFORMANCE CONCRETE MIX

Based on ACI 211.4 (93), high-strength concrete has a specified compressive strength fc′ of 42 N/mm^2 (6000 psi) or greater. This guide is intended to cover field strengths up to 84 N/mm^2 (12,000 psi) as a practical working range, although greater strengths may be obtained. Recommendations are based on current practice and information from contractors, concrete suppliers, and engineers involved in projects dealing with high-strength concrete.

5.11.1   REQUIRED STRENGTH

ACI 318 allows concrete mixtures to be proportioned based on field experience or laboratory trial batches. To meet the specified strength requirements, the concrete must be proportioned in such a manner that the average compressive strength results of field tests exceed the specified design compressive strength fc′ by an amount sufficiently high to make the probability of low tests small. When the concrete producer chooses to select high-strength concrete mixture proportions based upon field experience, it is recommended that the required average strength fc′ used as the basis for selection of concrete proportions be taken as the larger value calculated from the following:

- The average of all sets of three consecutive strength test results equals or exceeds the required fc′.
- No individual strength test (average of two cylinders) falls below 0.90 fc′. Note that this different from the ACI 318 requirement.

The latter criterion differs from the 3.4 MPa (500 psi) under strength criterion in ACI 318, because a deficiency of 3.4 MPa (500 psi) may not be significant when high-strength concrete is used.

High-strength concretes may continue to gain significant strength after the acceptance test especially if fly ash or ground granulated blast furnace slag is used.

Experience has shown that strength testing under ideal field conditions attains only 90% of the strength measured by tests performed under laboratory conditions. To assume that the average strength of field production concrete will equal the strength of a laboratory trial batch is not realistic, since many factors can influence the variability of strengths and strength measurements in the field. Initial use of a high strength concrete mixture in the field may require some adjustments in proportions and proper selection of its components. Once sufficient data have been generated from the job, mixture proportions should be reevaluated using ACI 214 and adjusted accordingly.

For high-strength concrete or high performance concrete there are no standard or typical mix proportions so it is more beneficial to present results on several successful mixes as presented by Neville (1983) in Table 5.22. This table contains different high performance concrete mixing ratios for different countries. There are nine mixes: (A) and (D) are from the United States; (B), (C), (E), (F), and (I) are from Canada; (G) is from Morocco, and (H) is from France.

TABLE 5.22
Mix Proportions for Some High Performance Concrete

Component (kg/m³)	Mixture								
	A	B	C	D	E	F	G	H	I
Portland cement	534	500	315	513	163	228	425	450	460
Silica fume	40	30	36	43	54	46	40	45	—
Fly-ash	59	—	137	—	—	182	—	—	—
GGBS	—	—	—	—	325	—	—	—	—
Fine aggregates	623	700	745	685	730	800	755	736	780
Coarse aggregates	1069	1100	1130	1080	1100	1110	1045	1118	1080
Total water	139	143	150	139	136	138	175ᵃ	143	138
w/c + b	0.22	0.27	0.31	0.25	0.25	0.30	0.38	0.29	0.30
Slump	255	—	—	—	200	220	230	230	110
Cylinder Strength (MPa)									
1	—	—	—	—	13	19	—	36	36
2	—	—	—	65	—	—	—	—	—
7	—	—	67	91	72	62	—	68	—
28	—	93	83	119	114	105	95	111	83
56	124	—	—	—	126	—	—	—	—
91	—	107	93	145	126	121	105	—	89
365	—	—	—	—	136	126	—	—	—

[a] It is suspected that the high water content was occasioned by a high ambient temperature in Morocco.

5.12 PUMPED CONCRETE MIX

5.12.1 BASIC CONSIDERATIONS

Concrete pumping is so established in most areas that most ready-mixed concrete producers can supply a mixture that will pump readily if they are informed of the pump volume and pressure capability, pipeline diameter, and horizontal and vertical distance to be pumped.

The shape of the coarse aggregate, whether angular or rounded, has an influence on the required mixture proportions, although both shapes can be pumped satisfactorily.

The angular pieces have a greater surface area per unit volume as compared with rounded pieces and thus require more mortar to coat the surface for pump ability.

5.12.2 COARSE AGGREGATE

The maximum size of angular or crushed coarse aggregate is limited to one third of the smallest inside diameter of the pump or pipeline. For well-rounded aggregate, the maximum size should be limited to two fifths of these diameters. The principles of proportioning are covered in ACI 211.1 and ACI 211.2.

Whereas the grading of sizes of coarse aggregate should meet the requirements of ASTM C33, it is important to recognize that the range between the upper and lower limits of this standard is broader than ACI 304 recommends for a pumpable concrete.

5.12.3 FINE AGGREGATE

The properties of the fine aggregate have a much more prominent role in the proportioning of pumpable mixtures than do those of the coarse aggregate.

Together with the cement and water, the fine aggregate provides the mortar or fluid that conveys the coarse aggregates in suspension, thus rendering a mixture pumpable.

Particular attention should be given to those portions passing the finer screen sizes based on Anderson (1977). At least 15% to 30% should pass the no. 50 screen and 5% to 10% should pass the no. 100 screen. ACI 211.1 states that for more workable concrete, which is sometimes required when placement is by pump, it may be desirable to reduce the estimated coarse aggregate content by up to 10% based on ACI 304R-30.

Exercise caution to ensure that the resulting slump, water-concrete mixture, and strength properties of the concrete meet applicable project specification requirements.

5.12.4 COMBINED NORMAL WEIGHT AGGREGATES

The combined coarse and fine aggregates occupy about 67% to 77% of the mixture volume. For gradation purposes, the fine and coarse aggregates should be considered as one even though they are usually proportioned separately.

ACI 304.2R includes an analysis worksheet for evaluating the pumpability of a concrete mixture by combining the fine and coarse aggregate with nominal maximum-sized aggregate from 3/4 to 1-1/2 in (19 to 38 mm). The worksheet makes provision for additional coarse and fine aggregate that can be added to improve the overall gradation and recognizes possible overlap of some coarse and fine aggregate components. If a mixture is known to be pumpable it is evaluated and graphed first; the curve representing its proportions provides a useful reference for determining the pumpability of a questionable mixture.

Those pumps with powered valves exert higher pressure on the concrete, and the most gradual and smallest reduction from concrete tube diameter can pump the most difficult mixtures. Concrete containing lightweight fine and coarse aggregate can be pumped if the aggregate is properly saturated. You can refer to ACI 304.2R for more detailed information and procedures.

5.12.5 WATER

Water requirements and slump control for pumpable normal weight concrete mixtures are interrelated and extremely important considerations. The amount of water used in a mixture will influence the strength and durability (for a given amount of cement) and will affect the slump or workability.

Mixing water requirements vary for different maximum sizes of aggregate as well as for different slumps. To establish the optimum slump resulting from water content for a pump mixture and to maintain control of that particular slump through the course of a job are both extremely important factors. Slumps from 2 to 6 in (50 to 150 mm) are most suitable for pumping. In mixtures with higher slump, the coarse aggregate can separate from the mortar and paste and can cause pipeline blockage. Slumps obtained using superplasticizers, however, are usually pumped without difficulty.

There are several reasons why the slump of concrete can change between initial mixing and final placement. If the slump at the end of the discharge hose can be maintained within specification limitations, it may be satisfactory for the concrete to enter the pump at a higher slump to compensate for slump loss, if the change is due simply to aggregate absorption.

5.12.6 CEMENTITIOUS MATERIALS

The determination of the cementitious materials content follows the same basic principles used for any concrete. In establishing the cement content, remember the need for over-strength proportioning in the laboratory to allow for field variations. The use of extra quantities of cementitious materials as the only means to correct pumping difficulties is shortsighted and uneconomical. Correcting any deficiencies in the aggregate gradation is more important.

5.12.7 ADMIXTURES

Any admixture that increases workability in both normal and lightweight concretes will usually improve pumpability. Admixtures used to improve pumpability include regular and high-range, water-reducing admixtures, air-entraining admixtures, and finely divided mineral admixtures.

Increased awareness of the need to incorporate entrained air in concrete to minimize freezing and thawing damage to structures has coincided with increased use of concrete pumps, as well as the development of longer placement booms. Considerable research and testing has established that the effectiveness of the air-entraining agent (AEA) in producing a beneficial air-void system depends on many factors. The more important factors are:

- The compatibility of the AEA and other admixtures and the order in which they are introduced into the batch
- The mixture proportions and aggregate gradation
- Mixing equipment and procedures
- Mixture temperatures
- Slump

AEA effectiveness and the resulting dosage depend on the cement fineness, cement factor, and water content, and the chemistry of cement and water, as well as that of other chemical and mineral admixtures used in the concrete. Refer to ACI 304.2R for more detailed information on air content and admixtures.

5.12.8   Field Practice

It is essential to perform preplanning for concrete pumping for successful placements, with increasing detail and coordination required as the size of the placement and the project increases.

This planning should provide for the correct amount and type of concrete for the pump used, provision for necessary pipeline, and agreement as to which personnel will provide the labor necessary to the complete placement operation.

Any trailer- or truck-mounted concrete pump can be used for pipeline concrete placement. The limiting factor in this method is the ability to spread the concrete as needed at the end of the pipeline. Generally, this is done by laborers using a rubber hose at the end of a rigid placement line.

The discharge of powered placement booms can be positioned at almost any point within the radius of the boom and at elevations achieved with the boom from near vertical (up or down) to horizontal. Boom use generally reduces the number of workers required for a given placement.

5.12.9   Field Control

Pumped concrete does not require any compromise in quality. A high level of quality control, however, should be maintained to ensure uniformity.

Concrete has been pumped successfully during both hot and cold weather. Precautions may be necessary to provide adequate protection during extreme conditions.

The concrete to be pumped must be well mixed before feeding it into the pump. Slump between 50 to 150 mm is generally recommended.

TABLE 5.23
Recommended Aggregate Grading

Size (mm)	Cumulative Percentage Passing	
	Max. Size 25 mm	Max. Size 20 mm
25	100	—
20	80–88	100
13	64–75	75–82
9.50	55–70	61–72
4.75	40–58	40–58
2.36	28–47	28–47
1.18	18–35	18–35
0.6	12–25	12–25
0.3	7–14	7–14
0.15	3–8	3–8
0.075	0	0

ACI 304.2R-91 recommends aggregate grading or pumping concrete as outlined in Table 5.23.

The research of Johansson and Tuutti in 1976 about pumping concrete revealed that the relation between the cement content and the void content determines suitability for pumping.

Generally, any mix selection of concrete to be pumped must be subjected to a test. Although laboratory pumps have been used to predict the pumpability of concrete, the performance of any given mix and the distance through which the concrete is to be pumped must be considered.

Figure 5.16 shows the relation between cement content and aggregate void content and effects of excessive frictional resistance on segregation and bleeding.

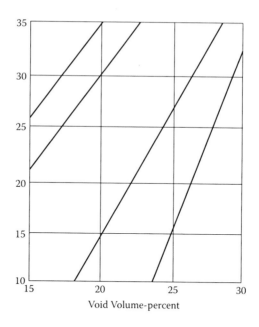

Void Volume-percent

FIGURE 5.16   Pumpability of concrete in relation to cement content and void content of aggregate.

REFERENCES

ACI 211.3-75, Revised 1987, Reapproved 1992. 1994. Standard practice for selecting proportions for no-slump concrete. *ACI manual of concrete practice, part 1: materials and general properties of concrete.* Detroit, MI: 19.

ACI 228-89-R1. In-place methods for determination of strength of concrete.

ACI Monograph No. 9. Testing hardened concrete: nondestructive methods.

Aitcin, P. C., and A. Neville. 1993. High-performance concrete demystified. *Concrete International* 15(1):21–26.

Anderson, W. G. 1977. Analyzing Concrete Mixtures for Pumpability, ACI JOURNAL, *Proceedings* V. 74, No. 9, Sept., pp. 447–451.

ASTM C31-91: Making and curing concrete test specimen in the fields.

ASTM C192-90a: Making and curing concrete test specimens in the laboratory.

ASTM C670-84: Testing of building materials.

Baaalbaki, M., et al. 1992. Properties and microstructure for high-performance concretes containing silica fume, slag, and fly ash. In *Fly-ash, silica fume, and natural pozzolans in concrete*, Vol. 2, ed. V. M. Malhotra, ACI, SP-132, 921–42. Detroit, MI.

Best, J. F., and R. O. Lane. 1980. Testing for optimum pump ability of concrete. *Concrete International* 2(10):9–17.

BS 812 Part 103-1985. Sampling and testing of mineral aggregate sands and fillers

BS 882:1992.

BS 1881. Testing concrete.

BS 1881:Part 6:1971. Methods of testing concrete, analysis of hardened concrete.

BS 1881:Part 108:1983. Method for making test cubes from fresh concrete.

Cadoret, G., and P. Richard. 1992. Full use of high performance concrete in building and public works. In *High performance concrete: from material to structure*, ed. Y. Malier, 379–411. London: E&FN Spon.

Causse, G., and S. Montents. 1992. The Roize Bridge. In *High performance concrete: from material to structure*, ed. Y. Malier, 525–36. London: E&FN Spon.

Johansson, A., and K. Tuutti. 1976. *Pumped concrete and pumping of concrete, CBI research reports*. Swedish Cement and Concrete Research Institute. 10:76.

Lessard, M., et al. 1994. High-performance concrete speeds reconstruction at McDonald's. *Concrete International* 16(9):47–50.

Neville, A. M. 1983. *Properties of concrete*. Pitman.

Part 5: Methods for testing hardened concrete for other than strength.

6 Special Concrete

6.1 INTRODUCTION

Technology developments and an increase in the use of reinforced concrete under different operating conditions has generated the need to use some materials to obtain some advantages in concrete characteristics to increase its strength or to protect the concrete along its lifetime.

Compressive strength is one of the most important properties as in most structural applications concrete is employed primarily to resist compressive stresses. All the codes consider the concrete's strength or its grade as the main factor from which can be known its other properties such as tension, shear strength, and the modulus of elasticity. For a long time it has been the dream of researchers and engineers to increase concrete's strength and to enhance its properties to make it a good competitor to steel structures.

Hot climate conditions affect the performance and durability of concrete, so there are some modern techniques and materials that should be used to overcome these issues.

In near future concrete will be considered as a durable material, but recently some problems and deterioration of concrete structures indicate that concrete is a durable material with some limitations. So it is important to use advanced materials that enhance and maintain concrete's durability.

6.2 ADMIXTURES

Now admixtures are widely used to increase the concrete's compressive strength, control the rate of hardening of fresh concrete, and increase the concrete's workability.

These admixtures are powder or liquid materials that are produced from carbohydrate, melamine, condensate, naphthalene, and organic and nonorganic materials.

The type of admixture should be identified correctly and also the dose that is required to achieve the target for the admixture.

There are some tests that should be applied to the admixture before use and results should be compared with the acceptance and refusal limits based on the project specifications.

The types of admixtures are as follows:

1. Normal setting water reducer increases the workability during concrete mixing without any change in water/cement ratio, or maintains the workability and decreases the water/cement ratio, so it will increase the concrete strength. This type matches ASTM C494, Type A–Normal Setting.

2. Retarder admixtures reduce the rate of reaction between cement and water to retard the concrete setting and hardening. This type of admixture follows ASTM C494, Types B and D and is usually used in ready mix concrete to achieve the required time to reach the site before setting the concrete.

3. Accelerators increase the rate of chemical reaction between cement and water, which increases the rate of setting and hardening concrete, and follow ASTM C494, Types C and E. Accelerators are not normally used in concrete unless early form removal is critical to the project's execution plan. Accelerators are added to increase the rate of early strength development or to shorten the time of setting. The advantages are as follows:

  i. Earlier removal of forms
  ii. Reduction of curing time
  iii. Early usage of the structure
  iv. Partial compensation for the effects of low temperature on rate of strength development
  v. Early finishing of surfaces
  vi. Reduction of pressure on forms

4. Admixtures for water reducers and retarders reduce water content by increasing the workability and retarding the setting time.

5. Admixtures for water reducers and accelerators reduce content by increasing the workability and accelerating the setting time; hence we obtain two functions by the same additive.

6. High range water reducers are much more effective than the admixtures discussed above. Based on Nevil (1983), at a given water/cement ratio, this dispersing action increases the workability of concrete, typically by increasing the slump from 75 to 200 mm, the mix remaining cohesive. Note that the improvement in workability is smaller at high temperatures.

7. High range water reducer and retarder additives increase the durability by reducing the water/cement ratio to increase the concrete compressive strength and increase the setting time.

6.2.1 Samples for Test

The admixtures exist in the form of liquid or powder, so the sample should be taken in the proper form to represent the deliverables quantity on site.

6.2.1.1 Powder Admixtures

The samples should be representative of one ton or less, and taken from six packs or from 1% of the total number of packs, or from all packs if the number of packs is fewer than six. The samples should represent all the packs.

6.2.1.2 Liquid Admixtures

The samples are taken from six drums or 1% of the total drums, whichever is larger, or from all the drums if the number is fewer than six. These samples represent an order, which contains not more than 5000 L of the liquid admixtures. Be sure to shake the drums to distribute the suspended materials and ignore the remaining sediments after shaking.

6.2.2   TESTS TO VERIFY ADMIXTURE REQUIREMENTS

6.2.2.1   Chemical Tests

The chemical test is performed to measure some parameters and compare them with the product data sheet to see if they match with manufacturer specifications or not.

1. Solid contents: For admixtures in the form of powder, the humidity will be removed from it by weight of about 3 g of the admixtures; remove the humidity and then note the percentage of the content of solid material.
2. For liquid admixtures: Pass 25–30 g of sand through sieve no. 30 into a glass bottle with rough surface opening and internal diameter 60 mm and height 30 mm. Do not cover. Put the bottle and the cover in the drying oven at 105–110°C and leave for 17 hours ± 15 minutes. Cover the bottle and place it in the dryer until it reaches room temperature and weigh it to the nearest 0.001 g and note the weight (W). Place about 4 mL from the sample inside the bottle over the sand and weigh it to the nearest 0.001 and assume it to be W_1. Put the bottle in the dryer oven at the same previous temperature and the same previous period, 17 hours + 15 minutes. Cover the bottle and place it in the dryer at room temperature and weigh it to nearest 0.001 g and assume it to be W_2. The percentage of the solid materials will be calculated from the following equation:

$$\text{Percentage of solid content} = (W - W_2)/(W - W_1) \times 100$$

6.2.2.2   Ash Content

The purpose of this test is to determine the content of nonorganic materials through analysis of ash content.

1. Heat the container with its cover at 600°C for 15–30 minutes and then transfer to dryer.
2. Leave it to cool for 30 minutes and then weigh it with the cover; the value of the weight is W_1.
3. Add about 1 g of the required admixture to the test and then re-cover and measure this weight value W_2.
4. To get less mechanical heat, spray the sample with a droplet of water and remove; then place it in the drying oven at 90°C.
5. Heat the sample to 300°C within an hour and then increase heat to 600°C for 2–3 hours, then leave the sample at 600°C for 16 ± 2 hours.
6. Transfer it to the oven. Dry it and allow it to cool with the cover in the dryer and weigh the contents to the nearest 0.001 g after 30 minutes of cooling; measure its weight value W_3.

The ash content value will be calculated from the following equation:

$$\% \text{ Ash content} = \frac{W_1 - W_3}{W_1 - W_2} \times 100$$

where

W_1 = weight of container + cover.

W_2 = weight of container + cover + sample weight before burning.

W_3 = weight of container + cover + sample weight after burning.

6.2.2.3   Relative Density

Use a sample at a temperature of $20 \pm 5°C$ and then transfer the sample to a tube with a capacity of 500 mL of the hydrometer immersed in the liquid inside the tube. Then lift the hydrometer to reach the balance and read the value of the hydrometer. Lift up to develop poise and read the grading at the base of the surface of contact with the liquid. Record the density to the nearest 0.002 gm/ml^3.

6.2.2.4   Define Hydrogen Number

A special apparatus will be used to define pH of a sample. Admixtures in the form of powders will be prepared in the form of liquid to determine pH and compare result with the specifications of the product.

6.2.2.5   Define Chloride Ion

Prepare two equal standard solutions of sodium chloride and then add each standard solution to the admixtures sample and estimate the proportion of chlorides after each addition. Calibrate it with a standard solution of silver nitrate to determine the point of volt differential.

We can assess the chloride ion content in the samples containing a very small percentage of chloride and at the same time estimate standard silver nitrate and control the quantity of required sodium chloride (Table 6.1).

Note that any method is permitted for determining the chloride content if it has the same accuracy.

TABLE 6.1

Admixture Characteristics

Character	Requirement
Solid content	Not more than 5% of weight for value stated by manufacturer for solid and liquid admixtures
Ash content	Not more than 1% of weight for value stated by manufacturer
pH level	Compare with number defined by the manufacture
Chloride ion content	Not more that 5% for value stated by manufacturer or 0.2% from weight, whatever is higher

6.2.3 Performance Test

These tests identify the performance of admixtures and the degree of influence on fresh and hardened concrete and compare the same concrete without admixtures. To determine the performance use two mixtures that have the same specification, one of which contains admixtures. The mixture without admixtures is called the control specimen. The admixture samples to be tested are obtained in a liquid or powder form as mentioned earlier.

6.2.3.1 Control Mixing

Control mixing utilizes ordinary Portland cement. The coarse aggregate must be identical to the standard specifications and be fully dry by using the oven and be clean and free of organic substances and impurities. The coarse aggregates have two sizes, 20–10 mm and 10–5 mm, and the flakiness index must not exceed 35%.

The sand follows the same specifications as the coarse aggregate and must also be dried by using the oven and have no more dissolved acid by 5%.

1. Ratio of specimen control without admixture:
   - Cement 300 ± 5 kg/m^3
   - Cement to whole aggregate ratio 1:6 by weight
   - Percentage of coarse aggregate by weight: 45% for sizes 10–20 mm, 20% for sizes 10–5 mm, and 35% for sand
   - Define the water quantity that provides slump 60 ± 1 mm or compaction ratio between 88% and 94%
   - Air entrained is not more than 3%
2. Ratio of specimen control with admixture:
   - Use the same ratios as in the case of mixing control, but add the admixtures with the same ratio as specified by the manufacturer. The water that makes the concrete mixture should not exceed the difference between the content of the air entrained in the control specimen by more than 2% and the total content of the air by not more than 3%.

Table 6.2 presents a comparison between the performance of the concrete with admixtures and the control without admixtures.

This table provides the guidance to accept or refuse the admixtures based on their function. For example, if we use a retarder time of only 30 minutes, based on the table the load should be refused as the retarding time must be from 1 to 3 hours.

The relation between admixtures and the minimum acceptable limits are shown in Table 6.3.

6.3 HOT WEATHER CONCRETE

Research about concrete mixing is done under moderate air temperature conditions that are about $27 \pm 2°C$ and most of the specifications recommend these circumstances. Often it will be the ideal temperature in the preparation of the concrete mixture for casting or curing.

TABLE 6.2
Performance Requirements for Concrete with Admixtures

Admixture Type		WR[a]	Retarder	Accelerator	WR and Retarder	WR and Accelerator	HWR[a]	HWR and Retarder
Max. water content as percentage of mix control		95	—	—	95	95	88	88
Air content		Not increased more than 25% in concrete with admixture than the control specimen without admixtures, and the total air content must not be more than 3%						
Hardening time at penetration resistance	0.5 N/mm²	During 1 hr from SC[b]	—	—	1 hr more than SC as lower limit	More than 1 hr	—	—
	3.5 N/mm²	During 1 hr from SC	—	—	—	1 hr lower than SC as lower limit	—	—
Setting time at penetration resistance[c]	3.5 N/mm²	During 1 hr from SC	From 1 to 3 hrs more than SC	From 1 to 3 hr lower than SC and not less than 45 min	From 1 to 3 hr more than SC	From 1 to 3 hr lower than SC and not less than 45 min	During 1 hr from SC[b]	From 1 to 3 hr more than SC
	27.6 N/mm²	During 1 hr from SC	Until 3 hr more than SC	Until 1 hr less than SC and not less than 45 min	Until 3 hr more than SC	Until 1 hr less than SC and not less than 45 min	During 1 hr from SC[b]	Until 3 hr from SC

a   WR is the water reducer and HWR is the higher water reducer.
b   SC is the specimen control mix.
c   Setting time according to ASTM C 494-1996.

TABLE 6.3

Relation of Admixtures and Minimum Limits to Concrete Strength as Percentage from Specimen Control Mix

Age (days)	WR	Retarder	Accelerator	WR and Retarder	WR and Accelerator	HWR	HWR and Retarder
1	—	—	125	—	125	140	125
3	110	90	125	110	125	125	125
7	110	90	100	110	110	115	115
28	110	90	100	110	110	110	110
180	100	90	90	100	100	100	100

Unfortunately, we cannot control nature as there are high temperatures in some countries and some countries have very hot weather in summer. Nowadays because of huge investment in real estate and competition, stopping work in a hot climate is not allowable.

Numerous tests have been conducted on concrete in the hot areas and there are particular specifications for hot regions by American Concrete Institute (ACI) Committee 305R, which has developed recommendations and requirements for hot weather.

In fact, the project contractor as well as the manufacturer of ready mix concrete follow recommendations; therefore, it is essential to define the cases where one must follow the requirements, precautions, and recommendations for concrete work in hot weather.

6.3.1 DEFINITION OF HOT WEATHER REGION

Hot weather must be defined in order to follow the instructions and guidance for concrete in hot weather. Several factors affect fresh and hardened concrete quality by quickening the setting time and loosening the moisture quickly. These factors are

- Higher weather temperature
- Higher concrete temperature
- Lower relative humidity
- High wind speed
- Sun rays

In a hot weather region, air temperature is continuously high, especially in the summer season, and the temperature is often higher than 35°C. In this case, all the precautions, recommendations, and instructions for hot climate regions should be considered.

The project specification document should define a hot climate as it varies from one country to another. Most project specifications require that the contractor must comply for concrete in hot climates if the weather temperature is equal to or higher than 35°C in the shade for three consecutive days. On the other hand, weather is considered ordinary when the temperature is less than 35°C in shade for three consecutive days.

It is customary to install thermometer in the location that represents the weather temperature on site clearly and accurately, and to calibrate the thermometer frequently to be confident of its accuracy.

6.3.2 PROBLEMS WITH CONCRETE IN HOT CLIMATE

Many problems can occur in a hot climate for both fresh and hardened concrete, and these problems are summarized below for better guidance on how to avoid these problems.

6.3.2.1 Fresh Concrete Problems

- Increase in the amount of water in the mix.
- The rate of concrete slump settlement will be decreased, which complicates the transportation operation, so more water is needed to increase slump settlement.
- The setting time will be decreased, which will be a problem in pouring, compaction, and finishing.
- The probability for plastic cracking increases.
- It is difficult to control air entrained inside the concrete.

6.3.2.2 Hardened Concrete Problems

High temperature at the site affects the performance of concrete after hardening, which is very dangerous since it impacts performance with age. In fact, this affects the economics of the project as a whole as will be discussed in the last chapter. The problems caused by high temperature:

- Decrease in strength at 28 days and after increasing the water or temperature of the concrete or both during concrete pouring or in the early days.
- Increase in the likelihood of shrinkage cracks as a result of the difference in temperature between the casting concrete and temperature of the building or different temperatures in the same sections.
- Concrete durability decreases with time due to these cracks.
- Color difference of the outer concrete as a result of various hydration rates or different water/cement ratios.
- A greater probability of corrosion for reinforcing bars due to concrete cracks.
- Concrete voids that increase the rate of permeability due excess water or poor concrete curing.

6.3.2.3 Problems Due to Other Factors in Hot Climate

There are other factors that cause problems when performing design or selecting materials to synchronize with hot weather in the working area:

- Using cement with a high dehydration rate
- Using a high strength concrete, which requires a large quantity of cement, thus causing problems in hot weather.
- A slender concrete section with high reinforcement ratio complicates compaction in hot climate.
- Economics of the project that necessitate working in high temperatures.

6.3.3 EFFECT OF HOT CLIMATE ON CONCRETE PROPERTIES

Concrete that has some distinguishing characteristics makes it a strong competitor with other materials used in construction, but those properties may be affected in a hot climate unless all the factors that affect the properties are under control. This control can be achieved by knowing the effect of hot temperatures on concrete properties.

In a hot climate, when the required precautions are not taken, a weak strength is obtained along with high permeability, which affects the structure performance over its life.

The factors that must be controlled are the appropriate choice of components of mixtures, defining the mixing ratios accurately, and performing the required tests before execution. We must also control the temperature of the concrete, wind speed, solar radiation, atmospheric temperature, and humidity during the stages of mixing, pouring, and curing.

Concrete that is mixed, poured, and cured in high atmospheric temperatures gives a higher compressive strength in the early days than in the case of low air temperature, but at 28 days and more it gives the least compressive strength result.

Figure 6.1 illustrates that with the increase in atmospheric temperature during the curing process, a reduction will occur in compressive strength at 28 days. The concrete strength is increased in first day based on research by Klieger (1958) and Verbeck and Helmuth (1968).

In low temperatures during mixing or curing it helps to dehydrate the cement, which significantly increases strength.

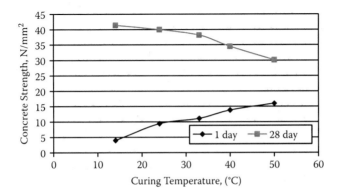

FIGURE 6.1   Effect of curing temperature on concrete strength.

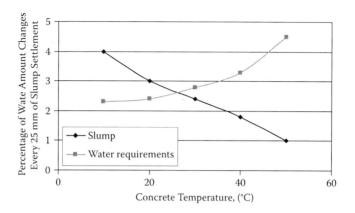

FIGURE 6.2   Effect of concrete temperature on concrete slump and water quantity.

Plastic shrinkage cracks happen more frequently in hot climates when pouring concrete for large surfaces, such as slabs, beams and foundations. These cracks are a result of drying the surface when the rate of evaporation of water from the concrete surface is greater than the rate of bleeding.

Evaporation increases due to an increase in atmospheric temperature, high wind speed, and low humidity. It can happen due to one factor only or more than one (Figure 6.2). The rate of bleeding depends on the components of the mixture, mixing ratios, thickness of the section, compaction, and process of surface finishing.

That is why the cracks of plastic shrinkage occur with increased evaporation of water from the surface. This is caused by the weather conditions or lack of bleeding rate, which is caused by the components of high strength concrete, in which silica fume or fly ash is used, or in the case of cement with very small particles in which the bleeding rate will be reduced. Therefore, in this type of mixing the probability for plastic shrinkage cracking is high even in normal weather conditions.

Note that plastic shrinkage cracking usually happens in hot climates. In Table 6.4, from ACI, different relative humidity, air temperatures, and concrete temperatures that cause high evaporation rates are presented. This table is based on a wind speed of 16 km/h and atmospheric temperature lower than the temperature of concrete by about 6°C.

Evaporation of water is not caused by the air temperature of concrete if other factors such as wind speed and humidity are present. Therefore, determining the maximum temperature of the atmosphere or of concrete may be correct in one region and not suitable for another area.

Therefore, for each project the weather temperature for a hot climate must be defined, and the maximum concrete temperature in general must be defined if the atmospheric temperature is higher than 20°C to 30°C, which requires experimental work in accordance with ASTM C192.

6.3.3.1   Control Water Temperature in Mixing

Water is an essential component of concrete and has an impact on the properties of fresh and hardening concrete.

TABLE 6.4

Temperature and Relative Humidity Levels Causing Plastic Shrinkage

Concrete Temperature (°C)	Air Temperature (°C)	Critical Evaporation Rate (kg/m²/hr)			
		1	0.75	0.5	0.25
		Relative Humidity (%)			
41	35	85	100	100	100
38	32	80	95	100	100
35	29	75	90	100	100
32	27	60	85	100	100
29	24	55	80	95	100
27	21	35	60	85	100
24	19	20	55	80	100

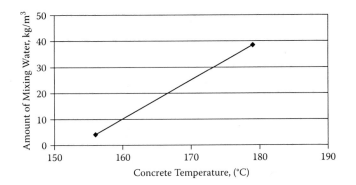

FIGURE 6.3 Effect of increasing concrete temperature on water amount and slump settlement 75 mm with maximum aggregate size 38 mm.

When increasing the temperature of the water, the concrete temperature will increase simultaneously. When increasing the temperature of concrete, the required water will increase to obtain the same slump, as illustrated in Figure 6.3.

The specific heat is the number of calories necessary to raise the temperature of one gram of the material one degree Celsius. The specific heat of water is about four to five times the specific heat of cement and aggregate.

Therefore, mixing water temperature has an impact on the temperature of concrete. Fortunately, the water temperature can be controlled easier than other components of concrete. The use of cooling water can reduce the temperature of concrete about 4.5°C. Note that cold water does not increase the quantity of water for mixing design. Reducing the temperature of water about 2°C to 2.2°C reduces the temperature of concrete about 0.5°C.

Contractors should obtain cold water and maintain the temperature during the entire process. So, the means of transporting the water and the water transportation systems such as the pipes, storage tanks, and transporting trucks and other transportation equipment should be insulated or painted white.

One can cool water up to 1°C by using ice or chilled water or other advanced technology.

Ice is frequently used in areas of hot weather. The ice should not be melted before mixing to obtain the most benefit but the ice must be melted at the end of the mixing process and before starting pouring.

The temperature of concrete can be estimated from the following equation:

$$T = \frac{0.22(T_a W_a - T_c W_c)T_w W_w + T_a W_{wa} + (-W_i)(79.6 - 0.5T_i)}{0.22(W_a + W_c) + W_w + W_i + W_{wa}}$$

where

T_a = aggregate temperature.
T_c = cement temperature.
T_w = temperature of mixing water without ice.
T_i = ice temperature.
W_a = weight of dry aggregate.
W_c = cement content.
W_w = quantity of water mixing.
W_{ma} = quantity of water absorbed by aggregate.

6.3.3.2   Control Cement Temperature

Increasing the concrete temperature will increase the dehydration rate, which will cause the concrete to set faster and lead to increasing the water to maintain workability. Increasing the water causes many problems, as discussed previously.

Some types of cement increase the setting time, and these types of cement are based on ASTM C595 type II Portland cement, or type IP or type IS.

Generally, in concrete that has a component that decreases setting time, the shrinkage cracking will be less.

We can use a type of cement that provides a slow rate of dehydration, which reduces the temperature caused by the dehydration—a process that reduces the overall concrete temperature. This will reduce the occurrence of cracks due to thermal expansion, and the likelihood of cracks will be less than with other methods to cool concrete. In general, using low heat cement is very important in mass concrete for slabs, retaining walls, and bridge supports.

The cement temperature affects the temperature of concrete as the concrete components constitute 10% to 15% of cement. That is why the temperature of concrete increases 0.5°C if the cement temperature increases about 4°C.

6.3.3.3   Control Aggregate Temperature

The aggregate is 60% to 80% of the concrete components and aggregate properties directly affect the concrete properties. The size, shape, and aggregate sieve analysis affect the amount of water added to achieve the required slump in the project specifications.

The aggregate grading and particle shape are important to reduce the amount of the water. Note that the aggregate from crushed stone needs more water and resists cracks better than round aggregate particles.

Reducing the aggregate temperature is very important to reduce the concrete temperature as the percentage of aggregate in concrete is high. Note that reducing the temperature of the aggregate from 0.8°C to 1.1°C reduces the temperature of concrete about 0.5°C. But reducing the temperature of the aggregate is costly. Some simple and cheap precautions to reduce the aggregate temperature or maintain its temperature at an acceptable limit include storing the aggregate in a shaded area or continuously spraying the aggregate with water.

6.3.3.4 Control Mixing Ratios

The design of the concrete mixture in a hot climate country must achieve the required specifications for strength after 28 days in accordance with ACI 318/318R, and a trial mixing must be performed based on the air temperature at the project location.

Suitable mixing design will be based on using the lowest cement content that provides the required strength and maintains performance over time.

In other words, replacing material such as fly ash or slag furnace can reduce the setting time, but will reduce the heat from the dehydration process. Moreover, different types of additives reduce the water in mixing, increase workability and reduce water content at the same time.

One can use a high water reducer to retard setting in the case of a long period between mixing and pouring and to obtain the reasonable workability to perform the compaction easily if it is preferred that the concrete slump be not less than 75 mm.

6.3.3.5 Control Concrete Mixing Process

Manufacturing reinforced concrete in a hot country needs special expertise of individuals who work at the site or in the factory. The production facilities should be able to supply concrete with the required specifications in hot weather conditions.

Note that the concrete temperature with the normal mixing ratio can be decreased by 0.5°C to:

- Reduce cement temperature by 4°C
- Reduce water temperature by 2°C
- Reduce the aggregate temperature by 1°C

The concrete mixer should be painted white to reduce the effect of the sun's heat and in some cases can be covered with a wet sheet.

The mixing speed must be lowered to reduce the heat produced by friction, as per ACI 207.4R. After the start of mixing and the mixture becomes homogeneous reduce the speed to one revolution per minute and the total number of revolutions around 300. The project specifications can define or modify this based on:

- The use of higher water reducer and retarder admixtures
- Using nitrogen to reduce the temperature
- If the concrete maintains its workability without adding water

FIGURE 6.4   Using nitrogen for cooling.

Nitrogen is used in some hot climate countries during the mixing process to reduce the concrete temperature. Figure 6.4 shows using nitrogen with a mixing truck.

Figure 6.5 shows the use of a block of ice in the concrete mix to reduce the concrete temperature. This is considered a cheap and effective technique to reduce the temperature so it is used widely in hot weather countries, especially in the Middle East.

6.3.3.6   Control Project Management

All personnel and equipment must be ready to deal with the concrete situation in a hot atmosphere and include experienced supervisors to control the site and solve any problem.

The period between the start of mixing and the start of pouring must be reduced. To achieve this there must be good management at the mixing center that delivers and the execution speed on site to avoid any delay in pouring or compaction.

It is preferable to choose a pouring time that is not close to the rush hour and prepare the site location carefully to allow trucks easy maneuverability on site without delay.

6.4   HIGH-STRENGTH CONCRETE

The concrete industry continues to develop concrete with greater compressive strength, and therefore every development redefines high-strength concrete.

FIGURE 6.5 Using ice in concrete mixing.

In the 1950s concrete compressive crushing strength at 28 days of 34 N/mm^2 defined high-strength concrete.

At the beginning of 1960 commercial concrete reached 41 and 52 N/mm^2 compressive strength. By 1970 the strength reached 62 N/mm^2.

Recently, compressive strength up to 138 N/mm^2.

Now, because of urban growth, especially vertical growth, use of high-strength concrete is significant. Moreover, some buildings have been difficult to construct without high strength, for example, the water tower in Chicago, and it is also used in bridges with large spans, such as the East Hington bridge that spans a river in Ohio.

In Australia, concrete that gives compressive strength of 65–70 N/mm^2 is widely used, and concrete that gives strength about 100 N/mm^2 is available as per Lioyd and Rangan (1995).

The evolution of the use of concrete is very important in economic terms as increasing strength allows smaller dimensions and this will reduce the dead load which will reduce the cost of the structure significantly. On the other hand, cities and other residential areas are in severe overcrowding. So any increase in space is a benefit to the owner in economic terms and this is what high-strength concrete provides since it reduces concrete, which increases the area and helps architects to achieve their goals easily.

High-strength concrete has special characteristics, as will be explained, which enhance concrete performance during the structure lifetime, and that will reflect economically by reducing the maintenance and the cost of regular inspection.

As a result of such developments the American Concrete Institute identified that 40 N/mm^2 or more is considered high-strength concrete, noting that this strength will not be from polymer or epoxy.

The committee also recommended that the definition of high-strength concrete varies according to location of the work site because of the nature of aggregate, for example, in Egypt, in the Hurghada and Aswan where they use crushed granite to achieve 45 N/mm^2 without any additions or enhancing mixing ratios. Under the same conditions using the aggregate from Cairo the same mix gives the highest value of 25 N/mm^2 compressive strength.

ACI mentioned that while strength for commercial concrete is 62 N/mm^2, high-strength concrete is defined as having strength from 83 to 100 N/mm^2. Practically, concrete strength is about 34 N/mm^2, but high-strength concrete exceeds 62 N/mm^2.

The Australian Society of Standards stipulates that compressive strength more than 50 N/mm^2 after 28 days is a high-strength concrete.

6.4.1 HIGH-STRENGTH CONCRETE COMPOSITION

6.4.1.1 Cement

High-strength concrete requires accuracy in choosing the components and is subject to quality control procedures. The cement is one of the basic components on which tests need to be conducted to identify strength to break at 7, 28, 56, and 91 days. According to the specifications of ASTM C109 do not use rapid hardening Portland cement Type III; the properties of cement and fines must follow the specification ASTM C150.

Increasing the cement content increases the temperature of concrete. For example, in the case of a 1.2 sq m column in a water tower containing 502 kg/m^3, cement reached temperature around 66°C although the initial temperature was 24°C during dehydration. These temperatures vanished within 6 days and there were no problems.

If problems are expected due to the rise in temperature, it is preferable to use low heat cement Type II to give the required strength after 28 days.

6.4.1.2 Mineral Admixture

The mineral admixtures such as silica fume, fly ash, and furnace slag form the basic material for high-strength concrete.

Admixtures are the main factors in developing high-strength concrete and self-compacting concrete. Now we need to ask some simple questions.

What is carrying the load aggregate or cement?
What is the function of the big aggregate?
What is the function of the sand?
What is the function of cement?

Simply put, the big aggregate is responsible for carrying the load on the concrete and the fine aggregate sand is used for filling the voids between the aggregates. Cement and water are used as adhesives between the big and fine aggregates. So to obtain high-strength concrete one needs to reduce the voids as much as possible.

When the voids are eliminated, the concrete can reach the highest strength and this is what the mineral additives such as silica fume, fly ash, and slag furnaces do, since they have small size grains that fill the voids in the concrete mixture.

The absence of the voids diminishes the force of water inside the concrete and also reduces the penetration of oxygen, and this prevents steel bars from corrosion, and the concrete will carry loads efficiently during the structure's lifetime without deterioration. Therefore, high-strength concrete in some references is called high performance concrete.

6.4.1.2.1 Silica Fume

Silica fume or micro silica is an important new material that is now widely used in the production of high-strength concrete. The silica fume is produced as a powder that is very fine and dense. This powder is made of rounded particles, 85% of which contain silicon dioxide. The average size of grains of silica fume is about 0.1 to 0.2 microns.

Figure 6.6 shows silica fume particles, and Figure 6.7 is a microscopic photo that presents the cement particle size on the left and the silica fume particle size on the right side. We can see how the silica fume particle is smaller than the cement particle.

Silica fume follows ASTM C1240 and this standard states the following:

- Content of SIO_2 is at least 85%
- The loss due to ignition is not more than 6%
- The maximum remaining percentage in a 45-micron sieve is 10%
- The minimum value of the surface area is 15 m^2/g

The smaller particle size as shown in Figure 6.8 will fill the voids in the concrete mix. Figure 6.9 presents cement particles on the left and the shape of mix after adding the silica fume as it fill the voids on the right.

This property is very important as using ultra fine powder in the concrete mix is the way to obtain durable concrete that works efficiently during its lifetime and the steel bars will not be affected by the weather conditions.

FIGURE 6.6 Silica fume particles.

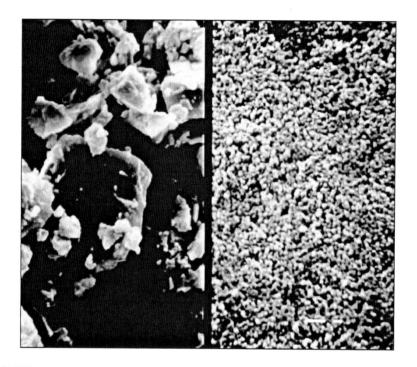

FIGURE 6.7   Size differences of silica fume and cement particles.

FIGURE 6.8   Shapes of silica fume particle.

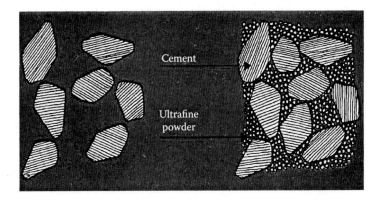

FIGURE 6.9   Shape of cement particles and silica fume.

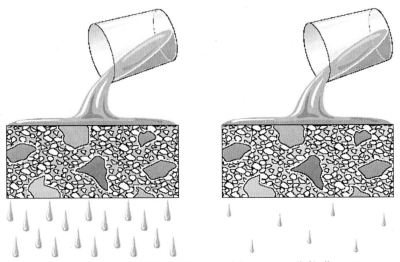

Schematic of concrete permeability. A high permeability concrete (left) allows water
to move into and through the concrete readily. Lowering the w/cm and adding silica fume can
reduce permeability to essentially zero. Such a reduction makes it very difficult for water
and aggressive chemicals such as chlorides or sulfates to enter the concrete.

FIGURE 6.10   Effect of using silica fume in concrete permeability.

In Figure 6.10 presents the pouring of water from above on a normal concrete slab
(left view) and one in which silica fume was used (right view). Note the difference
in the permeability of water. This figure clearly presents the benefit of using silica
fume to obtain high durability.

For any successful project, the supervisors should consider environmental impacts
and health of employees. Health, safety, and environment (HSE) are the main factors
for successful projects.

When using silica fume the laborers who work with this material must wear nose
covers as shown in Figure 6.11, as silica particles can affect the health.

Adding repulpable bags of densified silica fume directly to truck mixer for precautions regarding unopened bags. Note the use of a dust mask and safety glasses.

FIGURE 6.11   Precautions for adding silica fume to truck mixer.

Storage of silica fume on site must follow restricted procedures and precautions must be taken to store silica fume in silos as in Figure 6.12 or in completely closed sacks as shown in Figure 6.13. The correct way to open the sacks of silica fume is presented in Figure 6.14a and b.

The main disadvantage of the silica fume is pouring concrete in hot climate conditions, as the silica fume fills the voids, which prevents permeability. So the evaporation rate from the surface is higher than the bleeding so the curing process is very important in the case of using silica fume in concrete. Figure 6.15 presents the spraying of water to increase the humidity on the concrete surface.

Most specifications note that the shortest curing period is 7 days (Table 6.5). The curing process as shown in Figure 6.16 is performed by spraying chemical compounds during early curing until the concrete strength is increased and covering with wetting riprap or plastic sheet (Figure 6.17).

One consequence of the high early reactivity of silica fume is that the mix water is rapidly used up; in other words, self-desiccation takes place, as presented by Hooton in 1993, and at the same time the dense microstructure of the hydrated cement paste makes it difficult for water from outside, if available, to penetrate the unhydrated remnants of Portland cement or silica fume particles. In consequence, strength development increases much earlier than with Portland cement alone. Some experimental data are shown in Table 6.6 and indicate no increase in strength beyond 56 days. The data in Table 6.6 refer to mixes with a total content of cementitious materials of 400 kg/m^3, sulfate-resisting Portland (Type V) cement, silica fume contents of 10%, 15%, and 20% by mass of total cementitious materials, and a water/cement ratio of 0.36; the concrete specimens were maintained under moist conditions.

FIGURE 6.12   Silica fume storage in silos.

FIGURE 6.13   Storage of silica fume in sacks.

FIGURE 6.14   Putting sacks of silica fume in truck mixer.

FIGURE 6.15 Curing concrete with silica fume.

TABLE 6.5
Minimum Curing Times (in Days) Recommended in ENV

Rate of Gain of Strength	Rapid[a]			Medium			Slow		
Temperature of concrete	5	10	15	5	10	15	5	10	15
No sun, RH ≥ 80	2	2	1	3	3	2	3	3	2
Medium sun or medium wind or RH ≥ 50	4	3	2	6	4	3	8	5	4
Strong sun or high wind or RH < 50	4	3	2	8	6	5	10	8	5

Source: From Hooton, R. D. 1993. Influence of silica fume replacement of cement on physical properties and resistance to sulfate attack, freezing and thawing, and alkali-silica reactivity. *ACI Materials Journal* 90(2):143–151.

Note: RH, relative humidity in percent.

[a] Low water/cement ratio and rapid-hardening cement.

FIGURE 6.16 Spraying chemical compounds.

FIGURE 6.17 Curing by using plastic sheet.

TABLE 6.6

Strength Development of Test Cylinders of Concretes Containing Silica Fume

Age	Compressive Strength (MPa) of Mixes with a Silica Fume Content of (%)			
	0	**10**	**15**	**20**
1 day	26	25	28	27
7	45	60	63	65
28	56	71	75	74
56	64	74	76	73
91	63	78	73	74
182	73	73	71	78
1 year	79	77	70	80
2 years	86	82	71	82
3 years	88	90	85	88
5 years	86	80	67	70

TABLE 6.7

Typical Compressive Strength of Fly Ash Concrete

Cementitious Material	Compressive Strength, MPa at Age (Days)						
	1	**3**	**7**	**14**	**28**	**91**	**365**
Portland cement	12.1	21.2	28.6	33.9	40.1	46	51.2
Class F fly ash (25%)	7.1	13.9	19.4	24.3	30.3	39.8	47.3
Class C fly ash (25%)	8.9	19.0	24.1	28.5	29.4	40.5	45.6

6.4.1.2.2   Fly Ash

Materials with very fine particle size that fill the voids of the concrete to increase strength and durability (Table 6.7) are constantly sought.

One of the most successful materials is fly ash from coal-fired, electricity-generating power plants. These power plants grind coal to powder fineness before it is burned. Fly ash is the mineral residue produced by burning coal that is captured from exhaust gases and collected for use.

Fly ash particles are round (Figure 6.18) and their volume is less than the volume of cement particle so they fill voids in concrete mix and provide plasticity.

Additionally, when water is added to Portland cement, it creates a durable binder that glues concrete aggregates together and free lime. Fly ash reacts with this free lime to improve binding.

The spherical shape of fly ash creates a "ball bearing" effect in the mix, improving workability without increasing water requirements. Fly ash also improves the pumpability of concrete by making it more cohesive and less prone to segregation.

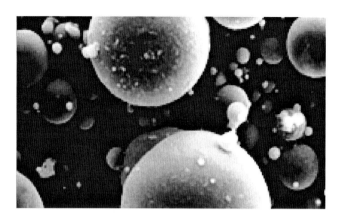

FIGURE 6.18   Shape of fly ash particle.

The spherical shape improves the pumpability by decreasing the friction between the concrete and the pump line. ASTM divides fly ash into two classes:

- Class F, normally produced by burning bituminous coal which has less than 5% CaO.
- Class C, normally produced by burning lignite or sub-bituminous coal; has CaO in excess of 10%.

A study by Gebler and Klieger in 1986 compared mix with 100% Portland cement and two other mixes of class C and F with 25% of the cement weight.

All the mixes had a total cementitious materials content of 307 kg/m^3 with a 25% content of fly ash by mass of total cementitious material. The water/cement ratio was 0.4 to 0.45, and the slump was 75 mm. The maximum size of the aggregate was 9.5 mm, and it the beneficial effect of the fly ash can be seen in that packing around the coarse aggregate is smaller than the conventional concrete mix.

From a durability point of view, fly ash prevents permeability; on the other hand, fly ash contains carbon so it is not recommended in cases of high probability of steel corrosion, and in some countries, as per the CUR report in 1991, it is prohibited in pre-stressed concrete.

6.4.1.2.3 Slag Furnace

Slag is another material used successfully to obtain high-strength concrete. Iron is manufactured using a blast furnace, and the furnace is continuously charged from the top with oxides, fluxing material, and fuel. Two products, slag and iron, are collected in the bottom of the hearth. Molten slag floats on top of the molten iron; both are tapped separately.

The molten iron is sent to the steel producing facility, while the molten slag is diverted to a granulator. This granulation process is the rapid quenching with water of the molten slag into a raw material called granules.

Rapid cooling prohibits the formation of crystals and forms glassy, nonmetallic, silicates and alumino silicates of calcium. These granules are dried and then ground

TABLE 6.8
Comparison of Chemical and Physical Characteristics of Portland Cement, Fly Ash, Slag Cement, and Silica Fume[a]

Property	Portland Cement	Class F Fly Ash	Class C Fly Ash	Slag Cement	Silica Fume
SiO_2 content (%)	21	52	35	35	85 to 97
AI_2O_3 content (%)	5	23	18	12	—
Fe_2O_3 content (%)	3	11	6	1	—
CaO content (%)	62	5	21	40	<1
Fineness as surface area m^2/kg[b]	370	420	420	400	15,000 to 30,000
Specific gravity	3.15	2.38	2.65	2.94	2.22
General use in concrete	Primary binder	Cement replacement	Cement replacement	Cement replacement	Property enhancer

Source: Sfa et al. (2002).

[a] Note that these are approximate values; values for a specific material may vary from what is shown.

[b] Surface area measurements for silica fume by nitrogen adsorption method; others by air permeability method (Blaine).

to a suitable fineness, the result of which is slag cement. The granules can also be incorporated in blended Portland cement.

6.4.1.2.4 Comparison of Mineral Additives

Every mineral additive has advantages and disadvantages and availability should be considered in the design mix.

Table 6.8 summarizes a comparison of cement, fly ash, slag, and silica fume. In general, the advantages of using mineral additives are:

- Better workability
- Easier finishability
- Higher compressive and flexural strengths
- Lower permeability
- Improved resistance to aggressive chemicals
- More consistent plastic and hardened properties
- Lighter color

Table 6.9 shows results from silica fume at different mixing ratios, and comparisons of compressive strength for every mixture until 56 days of age are shown in Figure 6.19. Table 6.10 presents different mixture ratios and the corresponding slump test values and compressive strength at 28 days.

6.5 SELF-COMPACTED CONCRETE

Beginning in 1983, the problem of the durability of concrete structures was a major interest in Japan. To make durable concrete structures, sufficient compaction by

TABLE 6.9

Strength Development of Mixtures Containing Silica Fume

Mixture	Cement (kg/m³)	Fly Ash (kg/m³)	SF (kg/m³)	SF (%)[a]	W/CM
1[b]	475	104	74	11	0.23
2[c]	390	71	48	9	0.37
3[b]	475	59	24	4	0.29
4[c]	390	—	27	6	0.35
5[c]	362	—	30	8	0.39
6[c]	390	—	30	7	0.37

[a] Silica fume as a percentage of total cementitious materials, by mass.
[b] Data from Burg and Ost (1994).
[c] Data provided by Elkem.

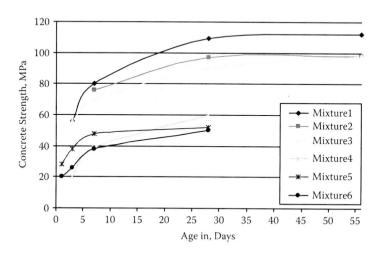

FIGURE 6.19   Strength development of several concrete mixtures containing silica fume.

TABLE 6.10

Types of Different Concrete Mixes

Concrete Composition	Mix 1	Mix 2	Mix 3	Mix 4	Mix 5
W/C	0.3	0.3	0.3	0.3	0.25
Water, kg/m³	127	128	129	131	128
Cement, kg/m³	450	425	365	228	168
SF, kg/m³	—	45	—	45	54
Fly ash, kg/m³	—	—	95	—	—
Slag, kg/m³	—	—	—	183	320
Coarse aggregate, kg/m³	1100	1110	1115	1100	1100
Fine aggregate, kg/m³	815	810	810	800	730
Superplasticizer, L/m³	15.3	14	13	12	13
Slump after 45 min., mm	110	180	170	220	210
Strength at 28 days, MPa	99	110	90	105	114

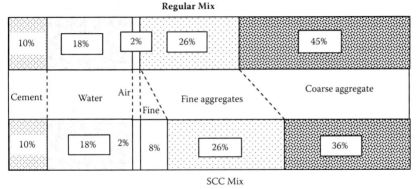

FIGURE 6.20 The difference between SCC and regular concrete mix.

skilled workers is required. However, the gradual reduction in the number of skilled workers in Japan's construction industry led to a similar reduction in the quality of construction work. One solution for the achievement of durable concrete structures independent of the quality of construction work is the employment of self-compacting concrete (SCC), which can be compacted into every corner of a formwork, purely by means of its own weight and without the need for vibrating compaction.

This type of concrete was proposed by Okamura in 1986. Studies to develop self-compacting concrete, including a fundamental study on the workability of concrete, were carried out by Ozawa and Maekawa at the University of Tokyo. Figure 6.20 shows the differences between SCC and traditional concrete mix.

6.5.1 DEVELOPMENT OF PROTOTYPE

The prototype of self-compacting concrete was completed in 1988 using materials already on the market. The prototype performed satisfactorily with regard to drying and hardening shrinkage, heat of hydration, denseness after hardening, and other properties. This mixture was named "high performance concrete."

At almost the same time, "high performance concrete" was defined as a concrete with high durability due to low water/cement ratio by Professor Aitcin. Since then, the term has been used around the world to refer to high durability concrete. Therefore, Okamura changed the name of the proposed concrete to "self-compacting high performance concrete."

6.5.2   Applications of Self-Compacting Concrete

Since the development of the prototype of self-compacting concrete in 1988, its use has gradually increased. The main reasons for the employment of self-compacting concrete include

1. Shorter construction period
2. Assuring compaction in the structure, especially in confined zones where vibrating compaction is difficult
3. Elimination of noise due to vibration, especially at concrete products plants

6.6   LIGHTWEIGHT AGGREGATE CONCRETE

The main disadvantage of normal concrete is the high self-weight as the density is about 2200 to 2600 kg/m^3. This high self-weight affects the dimensions for sections and foundations and has a direct impact on the project economics; therefore, there are many attempts to reduce the self-weight to increase efficiency. The practical range of densities of lightweight concrete is between about 300 and 1850 kg/m^3. Basically there is only one method for producing lightweight concrete by the inclusion of air and this can achieved in the following three ways:

1. By replacing the usual mineral aggregate with cellular porous or lightweight aggregate.
2. By introducing gas or air bubbles into the mortar to produce aerated concrete.
3. By producing no-fines concrete by omitting the sand fraction from the aggregate.

Table 6.11 presents the materials that are used in different ways.

Structural lightweight concrete has a density between 1350 and 1900 kg/m^3. As its name implies, this concrete is used for structural purposes and has a minimum compressive strength of 17 MPa (Table 6.12). Low-density concrete has a density between 300 and 800 kg/m^3. This concrete is used for non-structural purposes, mainly for thermal insulation. Its compressive strength, measured on standard cylinders, is between 7 and 17 MPa and the thermal insulation characteristics are between those of low-density concrete and structural lightweight concrete. For strength up to 41 MPa, ACI expresses the modulus of elasticity of concrete, E$_c$, in GPa as

$$E_c = 43 \times 10^{-6} \rho^{1.5} \sqrt{f_c'}$$

where

f_c = standard cylinder strength in MPa.

ρ = density of concrete in kg/m^3.

This expression is meant to be valid for values of density between 1440 and 2480 kg/m^3 but the actual modulus of elasticity may deviate from the calculated value by up to 20%.

TABLE 6.11
Lightweight Concretes

No-Fines Concrete	Lightweight Aggregate Concrete	Aerated Concrete	
		Chemical	Foaming Mixture
Gravel	Clinker	Aluminium powder	Preformed foam
Crushed stone	Foamed slag	Hydrogen peroxide and bleaching powder	Air-entrained foam
Coarse clinker	Expanded clay		
Sintered pulverized fuel ash	Expanded shale		
Expanded clay or shale	Expanded slate		
Expanded slate	Sintered pulverized fuel ash		
Foamed slag	Exfoliated vermiculite		
	Expanded perlite		
	Pumice		
	Organic aggregate		

TABLE 6.12
Strengths of Lightweight Aggregate Concrete and Cement Content

Compressive Strength (MPa)	Cement Content	
	With Lightweight Fine Aggregate (kg/m³)	With Normal Weight Fine Aggregate (kg/m³)
17	240–300	240–300
21	260–330	250–330
28	310–390	290–390
34	370–450	360–450
41	440–500	420–500

As far as lightweight aggregate concrete with a compressive strength in the range of 60 to 100 MPa is concerned, the relation of the modulus of the compressive strength seems to be best described by a Norwegian standard expression reported by Zhang and Gjorv (1991) as

$$E_c = 9.5f_c^{0.3} \times \left(\frac{\rho}{2400} \right)^{1.5}$$

where

E_c = modulus of elasticity in GPa.
f_c = compressive strength of 100 × 200 mm cylinders in MPa.
ρ = density of the concrete in kg/m^3.

The lower modulus of elasticity of lightweight aggregate concrete allows the development of a higher ultimate strain, compared with normal weight concrete of the same strength. The values of 3.3×10^{-3} to 4.6×10^{-3} have been reported.

6.6.1 Lightweight Aggregates

Lightweight aggregates have high porosity and low specific gravity. The main types of natural lightweight aggregates are diatomite, pumice, scoria, volcanic cinders, and tuff; except for diatomite, all are from volcanic origin. There are other types of lightweight aggregates known by a variety of trade names. The lightweight aggregates used in structural concrete are manufactured from natural materials such as expanded clay, shale, and slate.

6.6.2 Lightweight Coarse Aggregate or Structural Member

Requirements for lightweight aggregates are given in ASTM C330-89 and BS3797:1990. Table 6.13 presents the lightweight coarse aggregate requirements

TABLE 6.13
Grading Requirement of Lightweight Coarse Aggregate According to ASTM C330-89

Sieve Size	Percentage by Mass Passing Sieves Nominal Size of Graded Aggregate			
	25 to 4.75	19 to 4.75	12.5 to 4.75	9.5 to 2.36
25	95–100	100	—	—
19	—	90–100	100	—
12.5	2560	—	90–100	100
9.5	—	10–50	40–80	80–100
4.75	0–10	0–15	0–20	5–40
2.36	—	—	0–10	0–20

TABLE 6.14

Grading Requirement of Lightweight Coarse Aggregate According to BS3797:1990

| | Percentage by Mass Passing Sieves Nominal Size of Graded Aggregate | | |
Sieve Size	20 to 5	14 to 5	10 to 2.36
20	95–100	100	—
14	—	95–100	100
10	30–60	50–95	85–100
6.3	—	—	—
5.0	0–10	0–15	10–35
2.36	—	—	5–25

based on the ASTM. For manufactured aggregate the quality control on the grading is easier than for the natural aggregate. The lightweight coarse aggregate requirements based on BS3797:1990 are presented in Table 6.14.

6.6.3 LIGHTWEIGHT AGGREGATE CONCRETE DURABILITY

Lightweight aggregate concrete has a higher moisture movement than in the case of normal weight concrete. It has a high initial drying shrinkage about 5% to 40% higher than conventional concrete, but the total shrinkage with some lightweight aggregates may be even higher. Concretes made from expanded clay, shale, and expanded slag are in the lower shrinkage range.

Voids in the lightweight aggregate facilitate the diffusion of the carbon dioxide so increasing the cover thickness is recommended.

6.7 CELLULAR CONCRETE

In the initial classification of lightweight concrete, one method of reducing the density relied on the introduction of stable voids within the hardened cement paste or mortar. The voids can be produced by gas or by air; hence, the names gas concrete and aerated concrete. Because the air is introduced by a foaming agent, foamed concrete is another term. Strictly speaking, concrete is an inappropriate term because no coarse aggregate is present.

The introduction of gas is achieved usually by the use of finely divided aluminum powder, at the rate of about 0.2% by mass of cement. The reaction of the powder with alkalis liberates bubbles of hydrogen. The bubbles expand the cement paste or the mortar, which must have such consistency as to prevent their escape.

Cellular concrete may or may not contain aggregate, the latter generally is the case with concrete required for thermal insulation when an oven-dry density of 300 kg/m^3 and as low as 200 kg/m^3 can be obtained. When fine aggregate, either normal weight or lightweight, is included in the mix, the as-placed density lies between 800

TABLE 6.15
Guide to Cellular Concrete

Cement content, kg/m^3	300	320	360	400
As-placed density, kg/m^3	500	900	1300	1700
Oven dry density, kg/m^3	360	760	1180	1550
Fine aggregate content, kg/m^3	0	420	780	1130
Air content percent	78	62	45	28
Compressive strength, MPa	1	2	5	10
Thermal conductivity, Jm/m^{2}s $^\circ$C	0.1	0.2	0.4	0.6

TABLE 6.16
Typical Properties of No-Fines Concrete for Different Aggregate-to-Cement Ratios

Aggregate/Cement by Volume	w/c by Mass	Density (kg/m^3)	Compressive Strength at 28 Days (MPa)
6	0.38	2020	14
7	0.40	1970	12
8	0.41	1940	10
10	0.45	1870	7

and 2080 kg/m^3. Based on British Cement Association guidelines, Table 6.15 is guidance for cellular concrete.

6.8   NO-FINES CONCRETE

This is a type of lightweight concrete obtained when fine aggregate is omitted from the mix so it consists of cement, water, and coarse aggregate only. No-fines concrete is thus an agglomeration of coarse aggregate particles, each surrounded by a coating of cement paste up to 1.3 mm thick. Large voids exist within the body of the concrete and are responsible for its low strength, but their large size means that no capillary movement of water can take place.

The density of no-fines concrete in the case of normal aggregate varies between 1600 and 2000 kg/m^3, but in the case of lightweight aggregate the density is 640 kg/m^3. The compressive strength of no-fines concrete varies generally between 1.5 and 14 MPa, depending mainly on its density, which is governed by the cement content.

Table 6.16 presents the properties for no-fines concrete with aggregate size 9.5 to 19 mm as per McIntosh et al. in 1956.

6.9   SAWDUST CONCRETE

Sawdust concrete consists of roughly equal parts by volume of Portland cement, sand, and pine sawdust, with water to give a slump of 25 to 50 mm. It bonds well to ordinary concrete and is a good insulator.

Sawdust from tropical hardwood has been used to make sawdust concrete with a 28-day compressive strength of 30 MPa and a splitting strength of 2.5 MPa; the concrete had a density of 1490 kg/m^3.

Other wood waste, such as splinters and shavings, suitably treated chemically have also been used to make non–load-bearing concrete with a density of 800 to 1200 kg/m^3. Cork granules can also be used as per the study of Aziz et al. in 1979.

6.10   RICE HUSK ASH

Rice husk ash (RHA) is produced by burning rice husk in a controlled system to avoid environmental pollution, especially in countries where rice is a major plant like Japan and Egypt and other agricultural countries.

When it burns properly, it will produce high SiO_2 content and can be used as a concrete admixture. RHA exhibits high pozzolanic characteristics and contributes to the high strength and high impermeability of concrete. RHA consists mainly of amorphous silica for about 90% SiO_2, 5% carbon, and 2% others. The RHA specific surface area is between 40 and 100 m^2/g.

India produces around 122 million tons of rice every day. Each ton of paddy produces about 40 kg of RHA. There is a good potential to make use of RHA as a valuable pozzolanic material to produce the same results as micro silica.

In the United States, highly pozzolanic rice husk ash is patented under the trade name Agrosilica. It exhibits superpozzolanic properties when used in a small quantity and represents 10% by weight of cement and it greatly enhances the workability and impermeability of concrete.

6.11   ADVANCED MATERIALS FOR CONCRETE PROTECTION

The minimum requirements in various codes are often insufficient to ensure long-term durability of reinforced concrete in severe exposures such as those found in marine splash zones, bridges, and parking structures where de-icing salts are applied.

In addition, some of the newer structures (such as commercial buildings and condominiums) built in marine areas, but not in splash zones, are experiencing corrosion problems due to airborne chlorides. Marine structures in warmer climates such as in the Middle East, Singapore, Hong Kong, South Florida, etc., are especially vulnerable due to the high temperatures, which increase chloride ingress and the corrosion rate once the process is initiated.

In this section a brief description of supplemental corrosion protection measures is given for structures especially at risk. One of the most effective means to increase corrosion protection as described by El-Reedy in 2007 is to extend the time until chloride or a carbonation front reaches the steel reinforcement. The minimum code requirements allow the use of concrete with w/c ratio less than 0.45 and concrete cover thickness more than 38 mm—totally inadequate for the structures and environmental conditions outlined above if a design life of 40 or more years is specified. In many applications, designs complying with minimum code requirements would not provide as little as 10 years of repair-free service.

Precaution is better than cure. A protected concrete structure is easier and less expensive than one that needs repair. The reason is that the cost to repair and renovate reinforced concrete buildings is very high as in the case of foundation repair.

In reality, the protection of the structure from corrosion is the protection of the investment along the structure lifetime.

Recent developments provide economic and effective methods to protect steel reinforcement. Protection is very important to construction investments of billions of dollars worldwide.

The effective defense of steel bars is by external methods and these methods include using galvanized steel bars, epoxy-coated steel bars, stainless steel, or additives such as cathode inhibitor used during the pouring process, or an external membrane to prevent water permeability or provide cathode protection.

6.11.1 CORROSION INHIBITOR

There are two types of corrosion inhibitors. The first type is called the anodic inhibitor. The second type is the cathodic inhibitor. The anodic inhibitor depends on a passive protection layer on steel reinforcement. Cathodic protection is based on preventing the propagation of oxygen in the concrete. The most effective protection is the anodic inhibitor and it is also commonly used in practice.

Now we will explain these protections and clarify their advantages and disadvantages.

6.11.1.1 Anodic Inhibitors

The most common anodic inhibitor is calcium nitrate. It is compatible with pouring concrete at the site where there is no adverse impact on the properties of concrete, if it is in fresh or hardening state.

There are other types, such as sodium nitrate and potassium nitrate, which have high efficiency in the prevention of corrosion, but they are not used where there are existing aggregates with alkaline because they react with cement and cause extensive damage to the concrete.

Calcium nitrate has been widely used since the mid-1970s and its use accelerates the concrete setting time.

Broomfield (1995) suggests that retarders be added to concrete at the mixing plant to prevent damage. The mechanism that inhibits corrosion is related to stabilization of the passivation film; stabilization tends to be disrupted in the presence of chloride ions.

Anodic materials are used when concrete is directly exposed to chlorides such as in sea water. The corrosion inhibitor reacts with chlorides and increases the chloride concentration necessary to cause corrosion (see Table 6.17).

Determining the amount of calcium nitrate required based on the amount of chlorides to which the concrete will be exposed can be done in practice or through knowing the quantity of chlorides present.

To obtain high-strength concrete, choosing and constructing the right cover thickness and using the appropriate density concrete according to the specifications may

TABLE 6.17
Calcium Nitrate Required to Protect Steel
Reinforcement from Chloride Corrosion

Calcium Nitrate (kg/m^3 and 30% solution)	Chloride Ions on Steel (kg/m^3)
10	3.6
15	5.9
20	7.7
25	8.9
30	9.5

prevent the need for corrosion inhibitor for 20 years, but inhibitors are used when the structures are exposed to chlorides directly as in offshore structures.

6.11.1.2 Cathodic Inhibitor

Cathodic inhibitor is added to the concrete during mixing. Another type involves painting the concrete surface after hardening and transfer to the steel bars through the concrete's porosity and thus reducing the quantity of oxygen that propagates to the concrete. Oxygen is the important driver to the corrosion process.

There are many tests of corrosion inhibitors based on ASTM G109-92. These tests define the effect of the chemical additives on the corrosion of the steel reinforcement embedded in concrete. The corrosion inhibitor enhances the protection of the steel bars.

To obtain a higher efficiency from cathode protection it is essential to add a large quantity in the concrete mix to obtain more effective corrosion protection to the reinforced steel bars. On the other hand, these cathodic inhibitors, such as Ameen, phosphates, and zinc, retard the setting time when a large quantity is used and we must take the setting time into consideration.

We can conclude that the anodic inhibitor is more effective than the cathodic inhibitor. If we use the cathodic inhibitor, the increase of retarding time should influence our decision.

6.11.2 Coating of Steel Reinforcement by Epoxy

It is important to paint the steel bars using epoxies that are able to protect steel from corrosion. This method yielded positive results, especially in steel exposed to seawater, in a study performed by the Federal Highways Association (FHWA), which has been evaluating the use of epoxy to coat steel exposed to chloride attack.

Pike et al. (1994), Cairns (1992), and Satake et al. (1983) demonstrated the importance of painting steel reinforcement, and after 1970 epoxies were used in painting reinforced steel for bridges and offshore structures.

Some shortcomings must be avoided and precautions must be taken into account during the manufacturing and painting the steel, such as avoiding any friction between the bars, which would erode the coating layer. Also, it is difficult to use methods for measuring the corrosion rate such as polarization or half cell, and thus difficult to predict the steel corrosion performance or measure the corrosion rate.

Painting the steel reinforced bars is used extensively in the United States and Canada, and for more than 25 years coated bars have been used in more than 100,000 buildings, which is equal to two million tons of epoxy-coated bars.

The coated steel bar must follow ASTM A 775M/77M-93, which sets the following limits:

- The coating thickness in the range of 130–300 microns.
- Bending of the coated bar around a standard mandrel should not lead to formation of cracks in the epoxy coating.
- The number of pinhole defects should not be more than 6/m.
- The damage area on the bar should not exceed 2%.

These deficiencies cited by the code are the results of operation, transportation, and storage. There are some precautions that must be taken to avoid cracks in the paint as described in Andrade, Gustafson and Neff (1994), and Cairns (1992), which define the methods of storage and steel reinforcement bending, carrying the steel, and pouring concrete.

Painting steel reinforcement will reduce the bond between the concrete and steel, and therefore we must increase the length of steel bars to overcome this reduction in bond strength. According to ACI 318, the increase of the development length is about from 20% to 50%.

ACI 318 (1989) states, in the case of painting steel bars, the development length must be increased by 50% if concrete covers less than three times the steel bar diameter or the distance between the steel bars is less than six times the bar diameter, and in other circumstances increase the development length by 20%.

The Egyptian code did not take painting steel into account. In a study by El-Reedy et al. in 1995 the researchers found that the equation in the Egyptian code for the calculation of length can be applied for painting steel bars with epoxies without increasing the development length.

Painting mild steel is prohibited because the bond strength is due to friction, and paint removes all the bonding strength, so avoid in any way coating the smooth bars.

It is not recommended to increase the thickness of the painting coating; the thickness should not be more than 300 μm. When painting steel reinforcement with 350 μm to be used for steel reinforced in concrete slabs, testing found too many cracks, which led to the separation of steel bars and the concrete.

A comparison was done of steel bars without coating and bars coated with epoxy and exposing both to water from the tap and then placing samples of both in water containing sodium chloride and sodium sulfate. Then, the corrosion rates were compared and Table 6.18 shows the rates of corrosion for the coated and uncoated steel bars.

From this table one can see that the corrosion rate is very slow in the case of coated steel bars compared to the uncoated steel bars. This coating method is cheap and it is widely used by workers and contractors in North America and the Middle East.

TABLE 6.18

Corrosion Rates for Coated and Uncoated Steel Bars

| Case | Corrosion Rate (mm/yr) | |
	Tap Water	NaCl 1% + Na_2So_4 0.5%
Uncoated	0.0678	0.0980
Coated	0.0073	0.0130

However, note that the coating of reinforcement steel by epoxy is not exempt from requiring concrete of high quality while maintaining the reasonable concrete cover.

Some steel manufacturers can provide bars with the required coating. This is a very good option over performing coating on site where you cannot control the thickness of the coating because some special tools are needed to measure this thickness.

6.11.3 GALVANIZED STEEL BARS

Researchers in the United States recommend using galvanized steel in reinforced concrete structures. Moreover, the FHWA report recommends that the age of galvanized steel reach up to 15 years for high quality concrete under the influence of attacking chlorides according to the research of Andrade et al. in 1994. Galvanized bar is used effectively in structures exposed to carbon. Accelerated depletion of the galvanizing occurs if galvanized bars are mixed with ungalvanized bars.

Galvanization occurs through the use of a layer of zinc. The method is summarized as immersing a rod of steel into a zinc solution at 450°C and then cooling it. Hence, the zinc forms a cover on the steel bar. This cover consists of four layers: the outer layer is pure zinc and the other layers are a mix of zinc and steel.

Zinc is like most metal in that corrosion will happen with time. The rate of corrosion under different weather factors is calculated by the corrosion on the zinc layer and the time required.

The galvanized coating must be tested after bending the steel bars. The maximum zinc cover is around 200 microns based on ASTM A767/A767A M-90.

The stability of zinc essentially depends on the stability of the pH in concrete and pH equal to 13.3 is the value at which the passive protection layer is formed. But when the value of pH exceeds 13.3 the zinc will melt until it vanishes completely. Galvanization is totally dependent on the pH of the concrete pores and relies mainly on the alkalinity of Portland cement.

Hence, the shortage of cover in the event of galvanization at pH equal to 12.6 will be 2 μm, and in the case of pH equal to 13.2 shortage in the galvanized layer reaches 18 μm.

Laboratory tests have used different types of Portland cement with different alkalinity. If corrosion is equal, it is noted that when the thickness of the cover equals 60 μm, the lifetime is about 200 years in the case of a low alkalinity of cement and up to

11 years in the case of a high alkalinity of cement, as stated by the Building Research Establishment in the United Kingdom in 1969.

The thickness of the cover must be more than 20 μm in the American specification (ASTM 767/A767 M-90). This specification is for galvanized steel bars and identifies two types of galvanization, I and II, which have a cover thickness of more than 1070 and 610 g/m², respectively, which is the equivalent of 85 and 150 μm, respectively.

The maximum thickness of the cover of zinc is about 200 μm, which is recommended by the Building Research Establishment of the United Kingdom (1969); an increase in the thickness of the cover will reduce the bond between the steel and the concrete. When the pH is between 8 and 12.6 the zinc is more stable than if the pH value is increased, and note that the zinc gives high efficiency in the event that the structure exhibits significant carbonation as carbonation reduces the pH.

Galvanization will not work as corrosion prevention, but will significantly reduce the rate of corrosion. Galvanization can increase the value of chlorides, which accelerates corrosion from 150% to 200%. This will increase the time to cause corrosion by four times according to results from a study by Yeomans in 1994.

To study of the advantages and limits of galvanization, the important points according to the recommendation of the Concrete Institute of Australia in 1994 are

- Galvanization increases the protection of steel corrosion, but does not compensate for the use of concrete with a good cover.
- Avoid using galvanized steel with nongalvanized steel as this will cause corrosion quickly to the galvanic layer.
- Examine the galvanic layer after bending the steel reinforcement and fabrication and increase the bending diameter.
- Some precautions should be taken when using welding with galvanized steel.

6.11.4 STAINLESS STEEL

In some special applications reinforcing bars made from stainless steel are used to avoid corrosion. On the other hand, it has a higher cost so normal steel bars coated with a layer of stainless steel with thickness from 1 to 2 mm are used.

The same precautions for galvanized steel should be taken: do not use stainless steel or bars coated by stainless steel beside uncoated steel bars because it would lead to corrosion quickly.

In 1995, balconies using reinforcement stainless steel adjacent to normal steel underwent fast erosion, as stated by Miller.

The high cost of stainless steel makes its applications limited. The cost of stainless steel is about 8 to 10 times the cost of normal steel and higher than coating the steel bars with epoxy by about 15% to 50%.

The specifications for steel coated with stainless steel are under discussion by ASTM under the title Standard Specification for Deformed and Plain Clad Stainless Steel Carbon Steel Bars for Concrete Reinforcement and provides specifications for the installation of steel, which will follow specification A 480/480 M, for types 304, 316, or 316 L. Three levels of yield stress are 300, 420, and 520 MPa.

In the commercial market, there is reinforcing steel covered by stainless steel with yield strength about 500 MPa with maximum tensile stress of about 700 MPa, and the available diameters are 16, 19, 22, 25, and 32 mm, and now some factories are manufacturing 40 mm diameter.

It is not preferable to use stainless steel when welding, but if it is necessary, the welding will be done by tungsten inert gas and the welding wire will be the same material as stainless steel.

6.11.5 FIBER REINFORCEMENT BARS

In the last decade studies revealed theoretical and practical evidence for the replacement of steel reinforcement bars with fiber reinforced polymer (FRP) as this material is not affected by corrosion and will be very economical over a structure's lifetime.

In Figure 6.21, one can see the same shaped of rolled steel sections that are made from GFRP (glass reinforced polymer), which are more expensive than the traditional steel sections, but require no paint and no maintenance work as they are not affected by corrosion. Moreover, it has a light weight as the density of the fiber is 2.5 g/cm^3; the density of steel is about 7.8 g/cm^3, so GFRP weighs three times less than steel.

Note that because of its weight, which is very low, and its better resistance to corrosion than steel, it can be used for adding more floors to an existing building and replace the normal steel grating, handrails, and ladders in offshore structure platforms in the oil industry.

From the same materials and for the same reasons, bars have been developed as an alternative to the steel bars.

As can be seen in Figure 6.21, GFRP has protrusions on the surface, which gives it the ability to bond with the concrete. Now it is manufactured and used on a small scale and research focuses on the performance of these bars with time.

As a result of the limited production, the cost is very high, but the maintenance is very low. GFRP reduces the dead load to the structure, which will reduce the total cost of the structure.

Marine structures in Canada were built with bars of GFRP. A thick wall of precast concrete is designed to have maximum strength 450 kg/cm^2. The structures are exposed to temperatures of 35°C to –35°C.

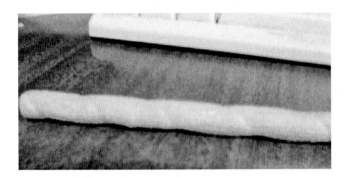

FIGURE 6.21 Shape of reinforcement bar from GFRP.

Bridges in Quebec, Canada, used this kind of reinforcement in areas of exposure to salt used to dissolve ice. One bridge was built in Antherino and another bridge on Vancouver Island. Vancouver performed tests on facilities exposed to different weather conditions. The tests were done through sampling and investigating reinforced concrete with an x-ray machine as well. These structures were aged between 8 and 5 years.

The research did not find impacts of various factors on the GFRP and the cycles of wet and dry did not affect the GFRP bars. By microscopy examination, a strong bond was found between the bars manufactured from GFRP and concrete.

The mechanical properties are summarized as the maximum strength about 5975 kg/cm² and the maximum bond strength 118 kg/cm², but the modulus of elasticity is less than the steel modulus of elasticity by about five times.

To overcome problems that might arise from creep, ACI 440 suggested that tensile stress must be not less than 20% less than the maximum tensile strength.

6.11.6   Cathodic Protection by Surface Painting

The corrosion of steel in concrete occurs as a result of attacking chlorides or carbonization of the surface and incursion in the concrete that reaches the steel, which reduces concrete alkalinity. Moisture and oxygen will produce complete deterioration of concrete.

There are some materials that are painted on the surface of the concrete to saturate the surface and penetrate the concrete to reach steel at speeds of 2.5 to 20 mm per day by either capillary rise phenomena, such as water movement, by penetrating with water as in the case of chloride attack, or through propagation by gas such as carbon dioxide when exposed to surface carbonation.

Therefore, when the materials reach the surface of steel they form an isolation layer around the steel bar surface that will reduce the oxygen to the surface on the cathode and reduces the melting of steel in the water in the anode area, thereby delaying corrosion and reducing its rate. Figure 6.22 presents the influence of this cathodic protection coating on the surface in protecting the steel bars.

These advanced materials are used in new construction or existing structures in which the steel bars have started corroding, or where corrosion clearly needs complete repair to the damaged concrete surface and then painting of the surface.

6.11.7   Cathodic Protection System

This is the most expensive method of protection, and it is usually used in protecting pipelines in the petroleum industry. It is intended for use in reinforced concrete structures and special structures due to its higher cost and need for special studies, design, execution, and monitoring.

Cathodic protection, if applied properly, can prevent corrosion of steel in concrete, and stop corrosion in progress. It accomplishes this by making the steel bar a cathode by use of an external anode. Electrons are supplied to the reinforcing bar from the anode, through the ionically conductive concrete. The current supplied

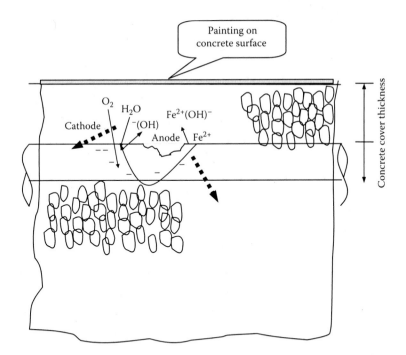

FIGURE 6.22   Painting concrete surface to provide cathodic protection.

should be sufficiently high so that all the local cells are inhibited and all the steel surfaces become anodic.

The external current can be supplied by connecting the steel to a metal that is higher in the electrochemical series (e.g., zinc). It serves as the anode relative to the cathodic steel. The anodes gradually dissolve as they oxidize and supply electrons to the cathodic steel. This type of cathodic protection requires a sacrificial anode.

An alternative method for cathodic protection is based on supplying electrons to the reinforcing steel from an external electrical power source. The electrical power is fed into an inert material, which serves as the anode, and is placed on the concrete surface. This method is referred to as impressed current anodic protection. The anode is frequently called a fixed anode.

6.11.7.1   Cathodic Protection

The principle of the use of sacrificial anode was discovered in 1824 by Sir Humphrey Davey and the method was used to protect the metal parts of submerged boats from corrosion.

At the beginning of the 20th century the technology was used for buried pipelines but when it was discovered that the soil is resistant to electricity, the use of cathodic protection with the current and constant direct current increased.

Cathodic protection is used in modern structures. The first practical application was in reinforced concrete on a bridge in a mountainous area in northern Italy.

Cathode protection became more common as a result of research and new technology. Broomfield surveyed consultants working on cathode protection projects

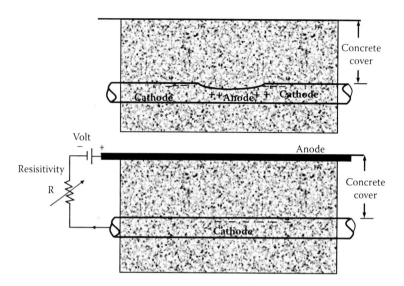

FIGURE 6.23 Cathodic protection method with sacrificial anode.

covering about 64,000 m² in the United Kingdom and the Middle East. The largest manufacturer supplied anodes that cover about 400,000 m² throughout the world and include garages and several types of bridges and tunnels. Cathode protection is used only for certain types of construction because of its high cost and the need for monitoring. It is used most often in structures exposed to environmental chlorides. Concrete corrosion results when chlorides weaken the concrete. Repair of concrete containing chlorides is considered impossible. Electrical protection effectively stops corrosion arising from pollution by chloride. Cathodic protection prevents or halts the corrosion of steel reinforcement in concrete. See Figure 6.23. Without cathodic protection corrosion of steel surfaces will develop.

The FHWA in the United States notes the only way to stop salt corrosion of concrete bridges is protection by electrical current regardless of the chloride content of the concrete. Electrons will generate on the steel surface and introduce anodes on the concrete surface. The concrete will become conductive. The cathodic protection is formulated on the steel reinforcement. The positive (sacrificial) anodes will move.

In this case the anode will made be from zinc and the corrosion will affect the zinc instead of the steel reinforcement as the oxidation will move to the steel reinforcement.

The second method of cathodic protection generates electrons on the steel reinforcement by using an outside source of electricity and a fixed anode.

Sacrificial protection is used in submerged structures. Because the concrete is immersed, electron movement is minimal and the potential voltage will be small and thus maintain cathodic protection for a long time. Figure 6.24 shows the method involving installation of an anode in concrete connected to an external electrical source.

An example of a fixed anode is wire mesh at the concrete surface that works as an anode. The conductivity, steel reinforcement, and batteries by using cables are shown in Figure 6.24.

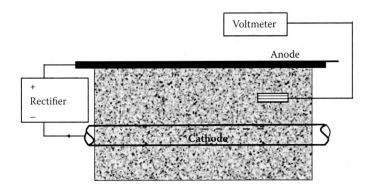

FIGURE 6.24   Cathodic protection method using external electricity.

The RILEM report suggested

- The electrical conductivity to the steel reinforcement must be continuous.
- The concrete between steel reinforcement and anode must conduct electricity.
- Alkaline aggregates must be avoided.

6.11.7.2   Cathodic Protection Components and Design Considerations

The cathode protection system consists of current, anode, and the electric conductive, which is the concrete. The concrete's relative humidity has a great effect on the electrical cables and the negative pole system, which is on the steel reinforcement. The last component of this system is the wiring between the anode and the source of direct current. Another important element is a control and measurement system.

The most important and expensive element is the anode as it should be capable of resisting the chemical, mechanical, and environmental conditions during the structure's lifetime. In general, its lifetime should be more than the coating layer's life.

Concrete structures supplied with cathodic protection are different from other applications because pores in some areas of concrete should not contain water; other areas may be dry or subject to water on their surfaces, unlike marine structures or structures buried in soil. If the concrete around the anode is completely dry, voltage should be increased to 10 to 15 instead of 1 to 5.

6.11.7.2.1   Source of Impressed Current

Most systems require a current about 10 to 20 milliamperes on the surface meter for steel reinforcement and take into consideration the lower layer of steel reinforcement. Current around 12 to 24 volts is always used to ensure little risk of causing electric shock to humans or animals.

The source of electric current is chosen in the design stage and the choice of electric source will be based on the fact that it must stop the corrosion process. A higher estimate is often assumed for electric current and 50% is added as a safety factor. This increase in current will produce little heat that will affect the strength of the current.

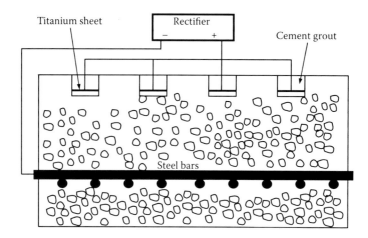

FIGURE 6.25   Anode holes at surface.

6.11.7.2.2   Anode System

The most important and expensive element in the cathodic protection system is the anode. Experience is needed in choosing and installing it. The choice of the anode system depends on the type of the structure and its shapes and other requirements.

There are two main types of anode. The first type is used on the bridge deck and it fixes on the deck from the top surface and needs special properties to accommodate vehicle movements and must be covered by an asphalt layer.

The second type is used in vertical buildings, and these anodes do not need high resistance to abrasion like the first type.

6.11.7.2.2.1 Anodes for Bridge Deck

There are two types. The first type is summarized as burying the anode in the concrete top layer. The main problem is that it increases the covering layer that will coat the anode, which consequently will increase the dead load on the structure. Therefore, it is not an economic solution.

The second type is executed by making holes on the concrete surface and installing the anodes inside them. This method can overcome the increase of dead load on the structure better than the first type. The difficulty is cutting parts of the concrete as the anodes should be adjacent to each other by around 300 mm to maintain the distribution of electricity in a suitable manner.

In the first type, carbon or silicon anodes with diameter 300 mm and thickness 10 mm are buried inside asphalt coke and the upper layer is an asphalt conductive layer as secondary anode.

The second type, shown in Figure 6.25, involves holes on the upper concrete surface with diameter 300 mm. The main target is to reduce the dead load on the bridge, which consequently will reduce overall construction cost and reduce the need for bridge surface leveling.

A titanium sheet can also be fixed on the bridge deck using plastic binder, as shown in Figure 6.26.

FIGURE 6.26   Fixing titanium mesh with plastic binder.

6.11.7.2.2.2 Vertical Surface Anodes   In this type of concrete structure member, titanium mesh is always used and often concrete is applied using shotcrete on the surface of the mesh to cover the anode. Caution is needed in using shotcrete as it needs competent workers, supervision, and contractor, and it also needs special curing to guarantee good quality concrete and good adhesive between it and the old concrete surface. There are some precautions that must be considered during construction to trust that the layer is saturated 100% and bonds with old concrete. The electric conductivity will be through titanium sheet welded on the mesh as shown in Figure 6.27.

FIGURE 6.27   Fixing titanium on concrete column.

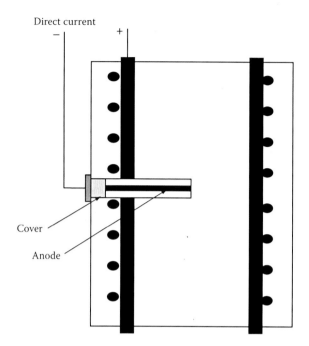

FIGURE 6.28   Buried anode.

In one method, we may use a rod of titanium platinum and place it inside the concrete member of a subject inside the coke so as to reduce the impact of acid generated on the anode. One of the most important precautions during implementation is to make sure that there is no possibility to cause a short circuit between the anode and steel reinforcement through the use of an appropriate cover that is placed after making the hole in the concrete, and will give warning of contact with steel bars, as in Figure 6.28.

The anodes should be distributed to ensure the protection of steel reinforcement in the structure, and this method is also used in horizontal surfaces.

6.11.7.2.3 Conductive Layer

The conductive layer is one of the most commercially successful. The anode is a main pole of inert metal and the secondary poles consist of layers from mortar, asphalt, and coating conductive to the electric current via carbon particles. Figure 6.29 shows the conductive layers in vertical reinforced concrete structures like columns and walls.

These layers are not durable over time but they are considered the cheapest and easiest way from a construction and maintenance point of view.

These layers can be painted to match the architectural design for the building. In addition, the method is reasonable for ceiling, vertical, and horizontal surfaces. The main disadvantage is that it cannot be used on surfaces exposed to abrasion as its capacity is low and its lifetime is about 5 to 10 years.

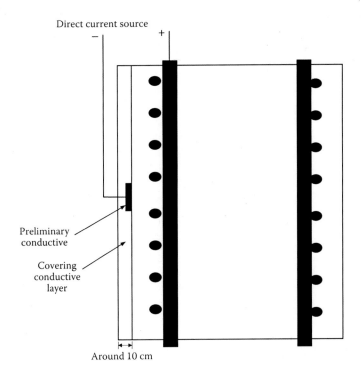

FIGURE 6.29   Anode conductive layer.

6.11.7.2.4   Cathodic Protection for Pre-stressed Concrete

Most specifications state that cathode protection must not be used with pre-stressed concrete. The movement of electrons results in the formation of hydrogen ions on the surface of steel reinforcement, and the presence of hydrogen affects the composition of iron atoms for high-strength steel, which is usually used in the process of pretensioning the steel for pre-stressed concrete.

Cathodic protection can be used, but hydrogen will not affect the nature of iron and the quantity of hydrogen resulting from the cathode protection process does not pose a threat to the pre-stressed steel.

There are some pre-stressed structures in which cathode protection is not used for the underground components. In some cases, sacrificial anodes are used in bridge piles as in Florida by using depleted anodes, as mentioned by Kessler et al. in 1995.

6.11.7.2.5   Bond Strength in Cathodic Protection

The impact of electric current on the bond strength between steel reinforcement and concrete is considered one of the most important subjects studied for several years. From a practical point of view there are no indications of the impact of cathodic protection on the bond strength.

Some structures that have been studied are about 20 years old, and there has been little impact on the bond strength as most of the bond strength is carried by the ribs in the deformed steel reinforcement.

TABLE 6.19

Advantages and Disadvantages of Different Protection Types

Method	Advantage	Disadvantage	Compatibility with Cathodic Protection
Increase concrete cover	No cost increase Obtain from good design	Has limit Increase cover will increase the shrinkage crack	Yes
Impermeable concrete	Depends on design	Need additives with very good curing	Yes, but silica fume must be <10% cement weight
Penetrating sealer	Economical	Complicated	Yes
Membrane prevents water	Well known technology	Can be damaged Difficult to fix without fault	Yes, but cathodic protection under membrane
Epoxy-coated steel bars	Known technology Little maintenance	Problem in quality control under certain weather conditions	Yes, with continuous electric current
Galvanized steel	Easy use on site	Fast damage by contact with non-galvanized steel	Yes, needs higher level of protection
Stainless steel	Excellent for corrosion prevention	High cost	Yes
Corrosion inhibitor	Calcium nitrate effect for little chloride concentration	Must know chloride amount Long-term effect not known	Yes
Cathodic protection	Known technology Long life of titanium anode	Regular inspection of rectifier	

Some practical experiences have proved that the bond strength is increased in the case of rust on the steel surface.

When using cathode protection there is no need to use another protection method, such as a membrane that prevents water permeability or painting the steel bars, thereby reducing the total cost of the project.

In some structures, due to the special nature of construction, it will be difficult to work the necessary repairs, so cathode protection is the best and only solution.

Table 6.19 is a summary of the methods of protection prepared by K. Keven and F. S. Daily in 1999, which gives the advantages and disadvantages of each method and the appropriateness of use with cathode protection at the same time.

REFERENCES

ACI 318-02, Building code requirements for structural concrete. In *ACI manual of concrete practice, part 3: use of concrete in buildings—design, specifications, and related topics*. 443.

ACI Committee 318. 1988. Revisions to Building Code Requirements for Reinforced Concrete. *American Concrete Institute Structures Journal* 85(6):645–74.

Andrade, C., J. D. Holst, U. Nurnberger, J. J. Whiteley, and N. Woodman. 1994. *Protection systems for reinforcement.* Prepared by Task Group VII/8 of Permanent Commisson VII CEB.

Anonymous. 1994. *The use of galvanized reinforcement in concrete, current practice note 17.* Concrete Institute of Australia.

Arnon B., S. Diamond, and N. S. Berke. 1997. *Steel corrosion in concrete: fundamental and civil engineering practice.* E&FN SPON.

Aziz, M. A., C. K. Murphy, and S. D. Ramaswamy. 1979. Lightweight concrete using cork granules. *International Journal of Lightweight Concrete* 1(1):29–33.

Bentur, A., S. Diamond, and N.S. Berke.1998. Steel Corrosion in Concrete: Fundamental in Civil Engineering Practice. London: E&FN SPON.

Broomfield, J. P. 1995. *Corrosion of steel in concrete: understanding, investigation and repair.* E&FN Spon.

Building Research Establishment. 1969. Zinc coated reinforcement for concrete. *BRE Digest* 109.

Burg, R.G., and Ost, B., 1994, Engineering Properties of Commercially Available High Strength Concrete, Research and Development Bulletin RD104T, Portland Cement Association, Skokie, IL.

Cairns, J. 1992. Design of concrete structures using fusion-bonded epoxy-coated reinforcement. *Proceedings of the Institution of Civil Engineers: Structures and Buildings* 4(2):93–102.

Clear, K., and Y. Virmani. 1983. Corrosion of non-specification epoxy-coated rebars in salty concrete. *Public Roads* 47(June).

Clifton, J. R., H. F. Beeghley, and R. G. Mathey. 1975. *Nonmetallic coatings for concrete reinforcing bars.* Building Science Series 65. US Department of Commerce, National Bureau of Standards. August.

CUR report 144: fly ash as addition to concrete. 1991. Gouda, the Netherlands: Center of Civil Engineering Research and Codes, 99.

El-Reedy, M. A., M. A. Sirag, and F. El-Hakim. 1995. Predicting bond strength of coated and uncoated steel bars using analytical model. MSc thesis. Faculty of Engineering, Cairo University.

El-Reedy, M.A. 2007. *Steel-reinforced concrete structure: assessment and repair of corrosion,* Boca Raton: FL: CRC Press.

Gebler, S. H., and P. Klieger. 1986. Effect of fly ash on physical properties of concrete. In *Fly ash, silica fume, slag, and natural pozzolans in concrete,* vol. 1, V. M. Malhotra, ed., 1–50. Detroit, Michigan: ACI SP-91.

Gustafson, D. P., and T. L. Neff. 1994. Epoxy-coated rebar, handled with care. *Concrete Construction* 39(4):356–69.

Hooton, R. D. 1993. Influence of silica fume replacement of cement on physical properties and resistance to sulfate attack, freezing and thawing, and alkali-silica reactivity. *ACI Materials Journal* 90(2):143–51.

Kendell, K., and F. S. Daily. 1999. Cathodic protection for new concrete. *Concrete International Magazine* 21(6).

Kessler, R. J., R. G. Powers, and I. R. Lasa. 1995. Update on sacrificial anode cathodic protection of steel reinforced concrete structures in sea water. *Corrosion 95,* Paper 516. Houston, TX: NACE International.

Klieger, P. 1958. Effect of Mixing and Curing Temperature on Concrete Strength, ACI JOURNAL, *Proceedings* V. 54, No. 12, June, pp. 1063–1081. Also, *Research Department Bulletin 103,* Portland Cement Association.

Lloyd, N. A., and Rangan, B. V., High-Strength Concrete Columns under Eccentric Compression, Research Report No. 1/95, School of Civil Engineering, Curtin University of Technology, Perth, Western Australia, January 1995.

McIntosh, R. H., J. D. Botton, and C. H. D. Muir. 1956. No-fines concrete as a structural material. *Proceedings of the Institution of Civil Engineers: Structures and Buildings.* Part I, 5(6):677–94.

Miller, J. B. 1994. Structural aspects of high powered electrochemical treatment of reinforced concrete. In *Corrosion protection of steel in concrete*, ed. R. N. Sawamy, 1400–514. Sheffield: Academic Press.

Okamura, H. and Ozawa,K. 1995. Mix design for self compacted concrete, *Concrete Library of JSCE*, 25, 107–120.

Ozawa K., M. Kunishima, K. Maekawa, and K. Ozawa. 1989. Development of high performance concrete based on the durability design of concrete structures. In *Proceedings of the second East-Asia and Pacific conference on structural engineering and construction (EASEC-2)*, Vol. 1:445–50.

Ozawa, K. and Maakawa, K. 1999. Development of SCC's Prototype Self Compacted High Performance Concrete, *Social System Institute*, 20–32.

Pike, R. G., et al. 1972. Nonmetallic coatings for concrete reinforcing bars. *Public Roads* 37(5):185–97.

Satake, J., M. Kamakura, K. Shirakawa, N. Mikami, and R. N. Swamy. 1983. Long-term resistance of epoxy-coated reinforcing bars. In *Corrosion of reinforcement in concrete construction*, A. P. Crane, ed., 357–77. United Kingdom: The Society of Chemical Industry/ Ellis Harwood Ltd.

Verbeck, G. J., and Helmuth, R. H., 1968, *Structure and Physical Properties of Cement Pastes*, *Proceedings*, Fifth International Symposium on the Chemistry of Cement, Tokyo, V. III, pp. 1–32.

Yeomans, S. R. 1994. Performance of black, galvanized and epoxy-coated reinforcing steels in chloride-contaminated concrete. *Corrosion* 50(1):72–81.

Zhang, M. H., and O. E. Gjorv. 1991. Mechanical properties of high strength lightweight concrete. *ACI Materials Journal* 88(3):240–47.

7 The Concrete Industry

7.1 INTRODUCTION

The aim of this chapter is to illustrate the traditional and modern techniques used to deliver high concrete quality and discuss recent standards. Nowadays, ready mix concrete is popular in some countries, but other countries use the mix on site. Based on a study by El-Reedy (2005), ready mix concrete provides high quality so it can be matched with less conservative design. Developed countries need more restricted codes to cover the uncertainty in concrete mix on site.

The concrete industry follows several steps. Every step has different alternatives, and selection of alternatives depends on the project and its economics. Every stage needs the same care to obtain good quality for the final product, and these steps are summarized as preparing the wood or steel forms, preparing and installing the steel bars, pouring concrete, compaction, and curing.

In this chapter, besides the traditional industrial process it is important to cover the alternative modern technique for every stage based on the American, British, and Egyptian standards.

7.2 EXECUTE WOODEN FORM

The design of wooden formwork depends on the structure system since each structure has a particular shape; for example, a solid slab and beam system are different from a flat slab with no beam and also different from a hollow block system.

Therefore, the design of the formwork is an important factor since the form will have to carry loads resulting from the concrete in its early stages during the casting and curing in addition to labor and equipment movement on the slab during pouring. Any error will result in many problems with the structure member which has been poured.

However, one can find that poor design or execution will result in many critical problems. As an example, a movement of the form in the beam side during concrete pouring will produce longitudinal cracks to the concrete, which affect the total structure integrity.

Based on ACI 318 there are some precautions in designing the form:

- Forms shall result in a final structure that conforms to shapes, lines, and dimensions of the members as required by the design drawings and specifications.
- Forms shall be substantial and sufficiently tight to prevent leakage of mortar.

- Forms shall be properly braced or tied together to maintain position and shape.
- Forms and their supports shall be designed so as not to damage previously placed structure.
- Design of formwork shall consider rate and method of placing concrete.
- Formwork design should be considered precisely, the construction loads, including vertical, horizontal, and impact loads.
- Special form requirements for construction of shells, folded plates, domes, architectural concrete, or similar types of elements.

In the following pages the wooden forms illustrated were provided by the International Network.

Figure 7.1 shows the wooden form of reinforced concrete wall, as well as all the vertical struts to support it. Figure 7.2 shows vertical, horizontal, and inclined struts from steel tubes and the concrete form is made from plywood to produce a smooth face to the bridge girder.

Figure 7.3 shows forms and struts and column boxes. Moreover, the wooden walkway is very clear to allow easy access and be safe for laborers' movement.

Figure 7.4 presents a wooden form for reinforced concrete slab that will be poured in the ground. The side shutter, strap, and strut are also shown.

Figure 7.5 shows the metal form for a big reinforced concrete foundation, and the metal is traditionally used in reinforced concrete pipes for sanitary or sewage work and also for pre-cast buildings.

In some special buildings steel sheet is used as a shuttering to the concrete slab and this shuttering remains and does not move after construction, as shown in Figure 7.6.

FIGURE 7.1   Formwork for reinforced concrete wall.

FIGURE 7.2   Bridges slab form support.

FIGURE 7.3   Wood form for concrete columns.

FIGURE 7.4   Strengthening wood form for ground slab.

Figure 7.7a–d illustrates another shape of wood form and support traditionally using plywood supported by steel supports. In Figure (7.7c) the cantilever is supported by steel to support the plywood form.

The use of wood or metal form must be conducted in accordance with the professional rules. The formwork design depends on the discretion of the designer who will approve the formwork officially.

The engineer who supervises the receipt of the forms should receive the formwork before the casting process and make sure it is good enough to avoid amendments so that the total time of the project is not affected.

The process of review and audits must be conducted through the supervisory documents to show that the form is sound. Be aware when designing the forms that they will be exposed to strong vibration during casting and compaction of concrete

FIGURE 7.5 Steel form for raft foundation.

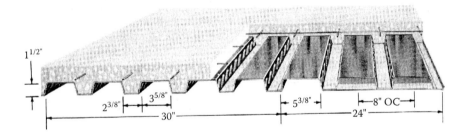

FIGURE 7.6 Steel sheet slab shuttering.

in addition to the movement of laborers, equipment, and tools. Following are some general observations that must be taken into account:

- The form should be solid and have strong provisions to prevent the leakage of mortar or the slurry, which is a mixture of cement and water, during the casting and compaction as this leakage is the main reason for honeycomb.
- If the wood or metal form is exposed to the sun for a long time before pouring it will be deformed. So the supervisor should make sure that there are no changes in the dimensions or torsion in the wood form.
- The bottom of the beam and slabs will camber according to the project specifications or the standard codes the project follows. In Egyptian codes the camber will be done for a span equal to or larger than 8 m from 1/300 to 1/500 from the span length; on the other side, for the cantilever that has a length higher than 1.5 m the camber will be 1/150 from the cantilever length.

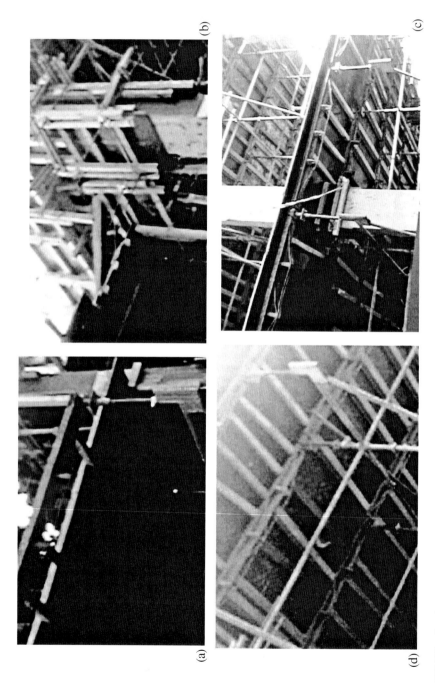

FIGURE 7.7 Steel support for counter form.

7.3 FORMWORK FOR HIGH-RISE BUILDINGS

Much of the change in concrete construction occurred in the first half of the 20th century. Advances in formwork, mixing, techniques for pumping, and types of admixtures to improve quality have all contributed to the ease of working with concrete in high-rise construction.

7.3.1 FORMWORK TYPES

The most efficient construction coordination plan for a tall building allows formwork to be reused multiple times. Traditionally, formwork was made of wood but as technology has advanced, the forms have become a combination of wood, steel, aluminum, fiberglass, and plastic, to name only a few materials. Each set may be self-supporting with trusses attached to the exterior or may need additional shoring to support it in appropriate locations. New additions to the family of forms include flying-forms, slip forms, and jump forms.

Flying forms or table forms are rental items. They are built in "typical" span lengths in order to provide continual reuse in a variety of jobs. The assemblage is made of fiberglass pan forms, steel trusses and purlins, and plywood, which are moved as a unit providing the base for a floor slab. After concrete placement and when it is determined that the strength of concrete has reached an appropriate maturity, the forms are removed, cleaned, and "flown" with a crane to the next level of a building for reuse (Figure 7.8).

Slip forms use materials that are continuously re-employed, as shown in Figure 7.9. Three types of jacks—hollow screw jack, hydraulic jack, and pneumatic jack—are used worldwide to "slip" formwork for a wall section to higher levels as the concrete cures. The screw jack is manually operated and used in areas where mechanization

FIGURE 7.8 Flying form.

FIGURE 7.9   Slip form.

is limited. The hydraulic and pneumatic jacks are fully automated, moving continuously as concrete is pumped into place.

Jump forms move as concrete cures to create a reusable, economic system. Jump forms also have a lifting mechanism but it is used differently from that of the continuous pour made with slip-forming. These are designed to swing away from the structure (like a door opening) for cleaning and oiling with subsequent reattaching to the wall as it increases in height.

Lift slab is another system, as shown in Figure 7.10a. This system involves pouring the slabs on the ground and lifting them by jack and anchor to the column as shown in Figure 7.10b.

7.3.2 Delivery Systems

Although the history of concrete is long, its use in tall buildings was partially curtailed because of difficult delivery systems. For the construction of the Ingalls Building, the ingredients of concrete were brought to the site and stored on the basement floor. Blending of materials was accomplished mechanically by power-driven, on-site mixers developed in the 1880s. When transport began in 1913, it was executed using

FIGURE 7.10   (Top) Lift slab. (Bottom) Anchoring lift slab.

open trucks. Since segregation occurred on the way to the site, remixing was always necessary. Effective means of transporting the quantities needed for such an enormous job required a transit-mix vehicle, which was not available until after 1920. In 1947, the first "hydraulically driven truck-mixers" were introduced to the scene.

Delivery of concrete had been an issue for tall buildings and other large projects. Another concern was the challenge of material placement in large quantities. Technology remained primitive until the 1960s when hydraulically powered and controlled pumps were first developed and mounted on trucks for mobile service. From here, techniques improved continually until now when pumping of concrete is considered even for small jobs.

7.3.3   ALLOWABLE TOLERANCE IN DIMENSIONS

Any project's specifications must contain tolerance in the dimensions to be followed by the laborers and the supervisor on site.

Table 7.1 illustrates the allowable tolerances in the dimensions in the Egyptian code in 2003, and Table 7.2 illustrates the allowable tolerances in the dimensions in ACI. These tolerance values are restricted to the site engineer to accept or refuse the work.

In the case of high-rise buildings there will be another specification for acceptance or refusal, as the designer must state the tolerance limits in the specifications of the project clearly for special structures.

7.4 DETAILING, FABRICATION, AND INSTALLATION OF STEEL BARS

There was a study performed by a research team to compile cases of collapse and cracks in reinforced concrete buildings. The team surveyed more than 150 cases and found among the reasons the lack of sufficient concrete cover to steel bars in beams and slabs during the pouring of concrete, which causes corrosion in the steel reinforcement and fall of concrete cover during installation of the steel bars on the forms. The site engineer who is supervising the work should take care to maintain the thickness of the concrete cover during the receipt of steel reinforcement work.

The steel reinforcement bars are very important elements in reinforced concrete projects as the location of the steel bars can change the state of a structure from full safety to a state of insecurity. The shape of the steel bars, their number, and their distribution locations are critical in designing reinforced concrete structures.

Therefore, the quality control team or the supervising engineer is fully responsible for ensuring that the steel bars in the drawings totally match those on the form before pouring, and any changes should be approved by the designer. This audit and review will be performed by the contractor QC team, and then be approved by the client representative.

The most common error is to move the steel bars during compaction and pouring. The steel bars must match the construction drawings and should be kept as they are.

Review of the drawings is a very important step and should be done through a checklist to avoid omissions. For example, with a weak review process you might see that there is not enough space between steel bars or not enough distance between steel bars and the form to pour the concrete without any honeycomb. A good reviewing system can easily avoid these mistakes.

Figure 7.11 shows the installation of reinforcing steel bars by using a crane due to the huge slab on the ground that must be cast.

Figure 7.12 illustrates the form of steel for a raft foundation. Figure 7.13 presents the sleeve and the shape of reinforcement around it. The shapes of reinforced steel bars that are connected to the concrete foundation as the retaining wall is seen are Figure 7.14. Figure 7.15 shows the arrangement of the steel reinforcement in a raft foundation.

7.4.1 Tolerance of Steel Bars in EGP

The tolerance of reinforcing steel depends on the shape of the bar itself. Figure 7.11 shows the most common forms in the steel reinforcement bars, and Tables 7.3 and 7.4 identify the allowable tolerances.

TABLE 7.1
Allowable Tolerances in Egyptian Code

Item	Description	Tolerance (mm)
1	Maximum tolerance to columns, beams and walls dimensions	
	In any span or every 6 m in any direction	±13
	Total structure dimensions	+25
2	Vertical alignment for column and walls surface and line of surface intersection	
	Every 6 m height	6
	Whole building height (max. 30 m)	25
3	Surface of corner column and vertical expansion joint	
	Every 6 m height	6
	Whole building height (max. 30 m)	15
4	Columns and walls execute by sliding form	
	Every 1.5 m height	3
	Every 15 m height	25
	Whole building height (max. 180 m)	75
5	Allowable tolerance for slab and beam bottom level	
	Every 3 m in horizontal distance	+5
	Every span or every 6 m horizontal distance	±10
	Whole building length and width	±20
6	Allowable tolerance in points of level that define slab and inclined beam leveling	
	Every span or every 6 m horizontal distance	±10
	Whole building length and width	±20
7	Allowable tolerance for columns, beam, slab, tie beam, and walls	
	Dimensions up to 400 mm	+10 or −5
	More than 400 mm	+15 or −10
8	Reinforced concrete foundation	
	Horizontal dimensions	−15 or +50
	Dimensions between axes	±50
	Foundation thickness	Without maximum or −2%
	Foundation top level	+15 or −5
9	Stairs	
	Height for one rise	±3
	Horizontal distance for one rise	±6
	Height for one flight or group of flights for one story	±5
	Horizontal dimensions for one flight or group of flights for one story	±10

TABLE 7.2

Allowable Tolerances in ACI Code

Item	Description	Tolerance (mm)
1	Maximum tolerance to columns, beams, and walls dimensions	
	In any span or every 6 m in any direction	+13
	Whole building dimensions	±25
2	Vertical alignment for column and walls surface and line of surface intersection	
	Every 3 m height	6
	Whole building height (max. 30 m)	25
3	Surface of corner column and vertical expansion joint	
	Every 6 m height	6
	Whole building height (max. 30 m)	13
4	Columns and walls execute by sliding form	
	Every 1.5 m height	3
	Every 15 m height	25
	Whole building height (max. 180 m)	75
5	Allowable tolerance for slab and beam bottom level	
	Every 3 m in horizontal distance	+6
	Every span or every 6 m horizontal distance	±10
	Whole building length and width	±19
6	Allowable tolerance in points of level that define slab and inclined beam leveling	
	Every span or every 6 m horizontal distance	±10
	Whole building length and width	±19
7	Allowable tolerance for columns, beam, slab, tie beam, and walls	
	Dimensions up to 304 mm	+10 or −6
	More than 304 mm	+13 or −10
8	Reinforced concrete foundation	
	Horizontal dimensions	−13 or +50
	Dimensions between axes	±50
	Eccentricity of column to the foundation	2% of foundation length in deviation direction and not more than 50 mm
	Foundation thickness	Without maximum or −2%
	Foundation top level	+13 or −50
9	Stairs	
	Height for one rise	±3
	Horizontal distance for one rise	±6
	Height for one flight or group of flights for one story	±3
	Horizontal dimensions for one flight or group of flights for one story	±6

FIGURE 7.11   Distributing steel bars on ground slab.

7.4.2   ALLOWABLE TOLERANCES IN ACI 318

In the American specifications, ACI states the allowable tolerances during steel bar fabrication or installation in the form and they are similar to the Egyptian code.

The allowable tolerances in the ACI are presented in Tables 7.5 and 7.6, and Figure 7.16. ACI states that the distance between bars higher than 25 mm (1 inch) in any case must be higher than three fourths the maximum nominal aggregate size.

7.5   CONCRETE COVER AND SPECIFICATIONS

The concrete cover is the first line of defense to protecting the steel reinforcement from corrosion, and therefore complete attention must be paid to the thickness of the concrete cover design and execution in all specifications. So, the thickness of the concrete cover must not exceed a certain limit, according to the types of the structure members and their surrounding environmental conditions.

FIGURE 7.12   Steel bars for raft foundation.

FIGURE 7.13   Wall steel bars with pipe sleeve.

FIGURE 7.14   Steel bars between foundation and wall.

FIGURE 7.15   Distributing steel reinforcement in concrete raft.

TABLE 7.3
Allowable Tolerances for Depth and Concrete Cover

Effective Depth	Tolerance in Effective Depth (mm)	Tolerance in Concrete Cover (mm)
Effective depth (d) ≤ 250 mm	±10	−6
Effective depth (d) > 250 mm	±15	−8

TABLE 7.4
Allowable Tolerances in Egyptian Code

Item	Member Description	Allowable Tolerance (mm)
1	Distance between bars	
	Beam	−5
	Slabs and walls	±20
	Stirrups	±20
2	Bending location and ends for longitudinal bars	
	Continuous beam and slabs	±25
	Ends of bars in beam and external slabs	±15
3	Decrease bar splice length	−25
4	Reduce splice length inside concrete	
	For bar diameters 10 to 32 mm	−25
	For bar diameters exceeding 32 mm	−50

TABLE 7.5
Allowable Tolerances for Depth and Concrete Cover

Effective Depth	Allowable in Effective Depth	Tolerance in Concrete Cover
Effective depth (d) ≤ 8 in	±0.375 in	−0.375 in
Effective depth (d) > 8 in	+0.5 in	−0.5 in

TABLE 7.6

Steel Bar Tolerances Based on ACI

Item	Member Type	Allowable Tolerance in Inches
1	Distance between bars	
	Min. distance between bars in beam	−0.25
	Slabs and walls in equal space	±2
	Stirrups	±1
2	Bending location and ends for longitudinal bars	
	Continuous beam and slabs	+2
	Ends of bars in beam and external slabs	+0.5
3	Reduce bar splice	−1.5
4	Reduce splice length inside concrete	
	For bar diameters #3 to #11	−1
	For bar diameters #14 to #18	−2

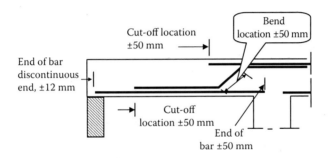

FIGURE 7.16 Tolerance in steel bars.

Therefore, many codes are devoted to the special specifications of the thickness of concrete cover, according to the nature of structure, method of construction, and quality of concrete used, as well as weather factors surrounding the structure.

7.5.1 BRITISH STANDARD

The British standard (BS 8110; 1985) considers the efficiency of the concrete cover to protect the reinforcement steel bars from corrosion as dependent on the thickness of the concrete cover and the quality of the concrete.

Moreover, when we talk about quality control of concrete and particular concrete cover, we mean that the concrete has no cracks and its compaction has been good since it will have a high density and a low ratio of water to cement (w/c), which works to prevent the permeability of the water into the concrete or any other material that will impact on steel reinforcement to cause corrosion.

Therefore, it is logical that concrete cover thickness is a function of the expected concrete quality as, for example, a high quality of concrete has little

TABLE 7.7

Properties and Thicknesses of Concrete Cover in BS8110

	Concrete Cover Thickness				
Concrete grade MPa	30	35	40	45	50
Water/cement ratio	0.65	0.65	0.55	0.50	0.45
Min. cement content, kg/m³	25	20	20	20	20
Environmental conditions:					
Moderate: concrete surface protected from external weather or hard condition	25	20	20	20	20
Moderate: concrete surface protected from rain or freezing and concrete under water or concrete adjacent to no affected soil		35	30	25	20
Hard: concrete surface exposed to rain and wetting and drying			40	30	25
Very hard: concrete exposed to seawater spray or melting ice by salt or freezing			50	40	30
Max. condition: concrete surface exposed to abrasion such as seawater containing solid particles, or moving water at pH 4.5 or machines or cars				60	50

need for cover thickness. This philosophy is loosely based on the British standard BS8110.

The specifications of the concrete cover thickness in the British code are indicated in Table 7.7, which depends on the weather factors to which the concrete structure is exposed, as well as the concrete strength and its quality based on the cement content and water/cement ratio (w/c).

7.5.2   AMERICAN CODE

The ACI code does not give the exact details, as does the British code. But the least thicknesses of the concrete cover in the case of concrete poured on site are shown in Table 7.8.

To get a good concrete exposed to water with high salinity ratio such as seawater, the American code sets the maximum ratio of w/c equal to 0.4, as well as less thickness of the concrete cover at 50 mm in the case of structures exposed to seawater.

Because of the mistakes expected during construction, it is preferable that the design of the concrete cover thickness be 65 mm so the least concrete cover will be 50 mm after execution.

The American code permits calculating the thickness of the concrete cover lower than what is shown in Table 7.8 in the case of pre-cast concrete.

7.5.3   EUROPEAN CODE

The European Union code gives precise and detailed recommendations, as well as defining the degree of concrete strength required based on the weather conditions faced by the structure.

TABLE 7.8

Minimum Cover Thickness for Cast-in-Place Concrete (ACI 301)

Type of Structure	Min. Cover (mm)
Concrete deposited against ground	75
Formed surfaces exposed to weather or in contact with ground	
No. 6 bar or greater	50
No. 5 bar or smaller	38
Formed surfaces not exposed to weather or not in contact with ground	
Beams, girders, and columns	38
Slabs and walls, no. 11 bar or smaller	19
Slabs and walls, no. 14 and 18 bars	38

The European code ENV-206 in 1992 set a ratio of w/c, as well as the lower content of cement in concrete and less thickness of the concrete cover corresponding to the concrete strength according to weather conditions. This is obvious from Table 7.9, which shows that all the choices are based on weather factors to which the structure is exposed. This code has set specifications for both reinforced concrete structures and pre-stressed concrete.

Note that the last column in the table, which determines the concrete strength characteristics, is set in MPa units. The first number is the value of the cubic strength characteristics and the second number is the equivalent of the cylinders, for example, C30/37 means concrete characteristics cube compressive strength of 30 MPa and concrete characteristic cylinder compressive strength is 37 MPa.

It is preferable to use cement resistance to sulfur if the sulfate content is more than 500 mg/kg in water or more than 3000 mg/kg in the soil. In both cases it is recommended to apply additional painting to the concrete surface.

Bear in mind that during construction the concrete cover thickness is often less than what has been designed because of the many factors during construction. This topic is discussed by Browne et al. (1983), who found the average thickness of the concrete cover to be about 13.9 mm, which is about half the value stated in the design (25 mm).

Van Daveer (1975) has done a survey on the thickness of the concrete cover in the design of bridges. If the design stated that the concrete cover thickness was 38 mm, it was found after some practical work that the standard deviation of the thickness of the concrete cover was very high, up to 9.5 mm. As a result, Arnon Bentur in 1997 suggested that if we are to obtain the thickness of concrete cover 50 mm at the site we must determine the thickness of the cover in design in the construction drawings and specifications of about 70 mm.

TABLE 7.9

Requirements for Concrete Durability According to European Code ENV-206 (1992) and British Specifications Recommendation DDENV 206

Exposure Condition	Max. w/c	Min. Cement Content (kg/m³)	Min. Concrete Cover (mm)	Conc. Grade
Dry	0.65	260	15	C30/37
Humid				
No frost	0.60	280	20	C30/37
Frost	0.55	280	25	C35/45
De-icing salts	0.5	300	40	C35/45
Seawater				
No frost	0.55	300	40	C35/45
Frost	0.50	300	40	C35/45
Aggressive chemical				
Slightly	0.55	280	25	
Moderately	0.50	300	30	
Highly	0.45	300	40	

The European code identified the previous deviation during construction; the minimum concrete cover must increase by the allowable deviation, and its value is from 0 to 5 mm in the case of pre-cast concrete and from 5 to 10 mm in the case of concrete cast in situ.

7.5.4 Special Specifications for Structures Exposed to Very Severe Conditions

Offshore structures are directly exposed to seawater, such as ports or offshore platforms or concrete used in the oil industry and located on the ocean. Based on the conditions of those structures specifications were set by ACI 357.

The concrete cover thickness is defined based on the construction method: whether it is a reinforced concrete structure or pre-stressed concrete. In addition, the concrete strength and the water/cement ratio limits are stated in this specification, as shown in Table 7.10.

The most dangerous corrosion area in the structure is the splash zone in that region exposed to seawater but not completely submerged in water and is not exposed to air only. So, there are several specifications of reinforced concrete in that type of region, as shown in Table 7.11.

In British specifications more detailed concrete cover thickness and concrete specifications have been identified for private structures. Moreover, the degree of mixing was also identified for all qualities of concrete and chloride diffusion factors in the concrete, and also in accordance with the life expectancy of the structure, as indicated in Table 7.12.

TABLE 7.10
ACI 357 Recommendation for Concrete Strength and Cover Thickness in Offshore Structures

			Cover Thickness	
Location	Max. w/c	Min. Concrete Strength at 28 Days	Reinforced Steel	Pre-stressed
Air	0.4	35	50	75
Splash zone	0.4	35	65	90
Immersed in water	0.45	35	50	75

TABLE 7.11
Comparison of Specifications for Concrete Design in Splash Zone

Code	Concrete Cover Thickness (mm)	Max. Crack Width (mm)	Max. w/c	Min. Cement Content (kg/m³)	Permeability Factor (m/s)
DNV	50	—	0.45	400	10–12
FIP	75	0.004× thickness or 0.3	0.45	400	—
BS6235	75	0.004× thickness or 0.3	0.40	400	—
ACI	65	—	0.40	360	—

The ability of the concrete cover to protect steel from corrosion does not depend only on the thickness of the cover, but also on the w/c ratio, the content of cement in the mix, as well as the degree of quality control of concrete. While those factors are the most important influences, corrosion protection also depends on the method of mixing, coarse aggregate and sand sieve analysis, the method of compaction, as well as curing of concrete after pouring.

7.5.5 EGYPTIAN CODE

The Egyptian code identified the thickness of the concrete cover that is required under environmental conditions surrounding the structure.

In the Egyptian code, the concrete cover thickness depends on the surface of concrete that has tension stresses and the effect of environmental factors, which have been divided into four types as shown in Table 7.13.

From this table one can accurately determine the structure for any type of condition and by knowing the strength of the concrete construction and the type of element one can determine the thickness of the concrete cover, as shown in Table 7.14.

The Egyptian code states that the thickness of the concrete cover should not be less than the largest bar diameter.

TABLE 7.12
British Code Requirements for Expected Diffusion Values, Expected Onset of Corrosion, and Corrosion Propagation

Source	Exposed Degree	Chloride Exposed Condition	Concrete Mix Detail					Age (Years)	
			Cement Content kg/m³	Max w/c	Min Slump (mm)	Min Concrete Cover (mm)	Diffusion Coefficient (m²/s)	Ci = 0.4%	Cp = 1%
All structures BS8110	Very severe	Spray sea water or de-icing	325	0.55	40	50	3.93×10^{-12}	3.1	5.6
			350	0.50	45	40	3.18×10^{-12}	2.6	4.6
			400	0.45	50	30	2.57×10^{-12}	1.9	3.7
	Severe	Abrasion and sea water contain solids	350	0.5	45	60	3.18×10^{-12}	5.8	10.4
			400	0.45	50	50	2.57×10^{-12}	5.4	10.2
Bridges BS5400 part 4	Very severe	De-icing or sea water spray	360	n/a	40	50	3.93×10^{-12}	3.3	6.1
			330	0.45	50	40	2.57×10^{-12}	3.0	5.5
	Severe	Abrasive by sea water	360	n/a	40	50	3.93×10^{-12}	5.5	10.3
			330	0.45	50	40	2.57×10^{-12}	5.7	10.3
Offshore structure BS 6349 Part1	Submerged always	Below sea level by 1m	350	0.5	n/a	>50 prefer 75	3.18×10^{-12}	3.3	7.2
							3.18×10^{-12}	7.4	16.2
	Tidal/splash	Less than lowest level by 1m	400	0.45	n/a	>50 prefer 75	2.57×10^{-12}	5.4	10.2
							2.57×10^{-12}	12.0	22.9
Sea water ENV06	XS1	Air saturated by water	330	0.5	40	35	3.93×10^{-12}	1.5	2.7
	XS2	Submerged in water	330	0.5	40	40	3.93×10^{-12}	2.0	3.6
	XS3	Spray or tidal	350	0.45	45	40	3.18×10^{-12}	2.1	4.6
Chlorides rather seawater	XS4	Wet and rarely dry	300	0.55	40	40	3.93×10^{-12}	2.0	3.4
	XS5	Wet and dry cycles	330	0.50	40	40	3.93×10^{-12}	2.0	3.6

Notes: Ci = Onset of corrosion, 0.4 wt% of cement. Cp = Propagation, 1 wt% of cement.

TABLE 7.13
Structure Element Based on Environmental Condition

Element Type	Degree of Exposed Tension Surface to Environmental Factor
Tension surface is protected	All internal members in building
	Element is always immersed in water lacking aggressive materials
	Last floor is isolated against humidity and rain
Tension surface is not protected	All structures direct to weather as bridges and last floor not isolated
	Structure tension surface is protected but near coast
	Element exposed to humidity, e.g., open hall or parking
Tension surface exposed to aggressive environment	Element exposed to higher relative humidity
	Element exposed periodically to relative humidity
	Water tanks
	Structure exposed to vapor, gases, or chemicals
Tension surface exposed to oxides that cause corrosion	Element exposed to chemical vapor cause corrosion
	Other tanks, sewage, and structure exposed to seawater

TABLE 7.14
Minimum Concrete Cover Thickness

	Concrete Cover Thickness (mm)			
	For All Elements Except Slab		Walls and Solid Slab	
Type of Element	$f_{cu} \leq 250$	$f_{cu} > 250$	$f_{cu} \leq 250$	$f_{cu} > 250$
Tension surface is protected	20	15	15	10
Tension surface is not protected	25	20	20	15
Tension surface is exposed to aggressive environment	30	25	25	20
Tension surface is exposed to oxides that cause corrosion	40	35	35	30

7.5.6 Execute Concrete Cover

From the previous sections, the importance of concrete cover became clear, and there are several practical methods to maintain the thickness of concrete cover during the construction process to match the design requirement.

The very famous method is to use pieces of cuboids of concrete called "biscuits" that are about 50 × 100 mm and their thickness depends on the required cover thickness as shown in Figure 7.11.

In these cuboids of concrete, we insert a steel wire during pouring to tie the steel bars with these pieces to maintain the spacing between the bars as in the drawings.

The disadvantage of this method is that when the workers move along the steel bars to pour concrete or for inspection, supervision, and other activities, this load will be a concentrated pressure on the concrete pieces and will lead to cracks and damage. Therefore, after construction the cover thickness is very small or has vanished.

The advantages of this method are that it is low cost since pouring it is the same as mixing on site and the workers are available on site also.

Another method that is more practical is to use plastic pieces that will maintain the concrete cover. This will cost little money, and provides the thickness of concrete cover required by the drawings of the designer.

Recently, many contractors have been using these parts of plastic, which are cheap and strong maintain cover accuracy. Figure 7.17 shows different types of plastic pieces based on the bars' diameters and the location of the steel bars.

These plastic forms are used to stabilize the steel bars and to maintain the thickness of the concrete cover for slabs, beams, columns, or foundations. The plastic covers vary according to the size and form of bars and concrete cover thickness to be preserved.

Therefore, the customer must identify for suppliers all the required information on the quantity and thickness of the concrete cover, the diameters of the steel bars that will be used, and the type of the concrete member that will be installed as these pieces of plastic differ depending on whether they are used in a column, slab, or beam.

Figure 7.17 shows plastic for the intersections of bars, and it is worth mentioning that these kind of biscuits fix the steel bars and prevent them from moving, which also maintains the distance between steel bars.

In Figure 7.18, one can see the plastic pieces carrying the chair, which are useful to protect steel that is exposed to the concrete's outer surface.

In the past 30 years another method that was used to keep the concrete cover is to put reasonable aggregates underneath the steel bars, but this is a shabby method and is prohibited. It is important to highlight that this method is still used in some low cost residential buildings in some developed countries but will prove more expensive because of deterioration of the building due to corrosion effect and cost of performing the repairs.

After construction, it is recommended in some special structures that are exposed to severe conditions to highlight in the specifications to measure the concrete thickness.

The British standard requires that the measurement of concrete cover shall be carried out in accordance with BS 1881, Part 204 using an electromagnetic device that estimates the position depth and the size of reinforcement. The locations for checking cover and the spacing between measurements shall be advised by the engineer based on the objective of the investigation. Reinforcement should be secured against displacement outside the specified limits unless specified otherwise:

1. The concrete cover should be not less than the required nominal cover minus 5 mm.

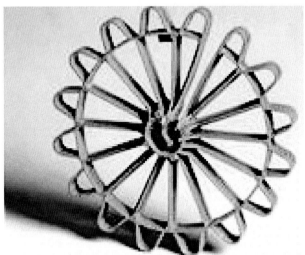

FIGURE 7.17 Plastic pieces. *Continued.*

FIGURE 7.17 *Continued.*

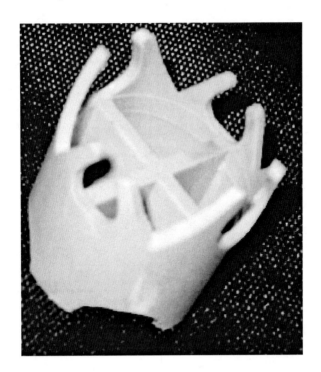

FIGURE 7.17 *Continued.*

FIGURE 7.18 Plastic pieces carrying chairs.

2. Where reinforcement is located in relation to only one face of a member, for example, a straight bar in a slab, the concrete cover should be not more than the required nominal cover plus:

   a. 5 mm on bars up to and including 12 mm size
   b. 10 mm on bars over 12 mm up to and including 25 mm size
   c. 15 mm on bars over 25 mm size

Nominal cover should be specified for all steel reinforcement including links. Spacers between the links (or the bars where no links exist) and the formwork should be of the same size as the nominal cover.

Spacers, chairs, and other supports detailed on drawings, together with such other supports as may be necessary, should be used to maintain the specified nominal cover to the steel reinforcement. Spacers and chairs should be placed in accordance with the requirements of BS 7973-2.

Spacers and/or chairs should conform to BS 7973-1. Concrete spacer blocks made on the construction site should not be used.

The position of reinforcement should be checked before and during concreting, particular attention being directed to ensuring that the nominal cover is maintained within the given limits, especially for cantilever sections. The importance of cover in relation to durability justifies the regular use of a cover meter to check the position of the reinforcement in the hardened concrete.

7.6   CONCRETE POURING

A critical process that needs more concentration and competent persons on site is the concrete pouring. This process should be done correctly and avoid movement of the steel bars.

All the codes and standards provide constraints to avoid concrete segregation, and the time between adding water to dry mix and pouring must not be more than 30 minutes in normal weather conditions where the temperature does not exceed 30°C in the shade and 20 minutes in the case of hot weather, and this time period can be increased if needed by using the admixtures listed in the project specifications.

If it is essential that the pouring be done in a special situation such as for high columns and retaining walls, for example, it should be cast in layers with thickness 300–500 mm accompanied using a vibrator. Note that the interval between each layer should not exceed 30 minutes in normal weather or 20 minutes in hot weather. These periods can be extended by adding admixtures already approved when there is steel reinforcement enough to bond subsequent casting layers.

Most columns span more than 2.5 m and allow pouring in one step. But the side of the form will be every 25 m height maximum to pour concrete from it and close the opening and open the above one and so on until finishing the column height.

ACI presents the right and wrong ways to pour concrete to avoid segregation and honeycomb in the column or the walls. Figure 7.19 presents the proper way to pour the concrete in columns and retaining walls. Figure 7.20 shows the proper way to pour a ground slab. Figure 7.21 illustrates the proper way to pour the concrete for inclined slabs to avoid the segregation of the aggregate. Figures 7.22 and 7.23 present

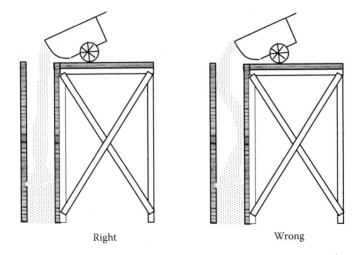

FIGURE 7.19 Pouring concrete wall.

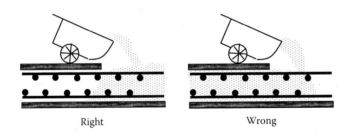

FIGURE 7.20 Pouring ground slab.

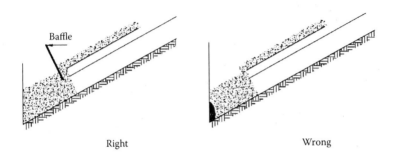

FIGURE 7.21 Pouring inclined concrete slab.

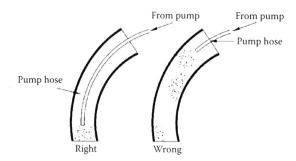

FIGURE 7.22   Pouring concrete in curvature of wall.

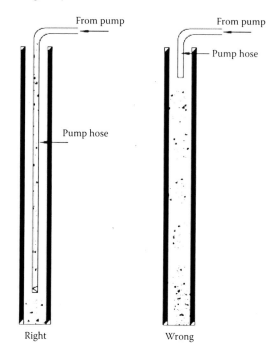

FIGURE 7.23   Pouring concrete by pump.

the right way for pouring the concrete by pump (Figures 7.24–7.26) for a curved beam or wall and the retaining walls or columns, respectively.

Environmental conditions must be taken into consideration during concrete pouring as in hot climates with temperature more than 36°C in shade during mixing or pouring, the big aggregate and sand should be stored in a shaded area and the aggregate can be cooled with water. For loose cement in silos, the silos must be painted with a material and color that reflect the sun from outside. Cement sacks should be put under a slab or in a shaded area. The water can be cooled during pouring. In the case of pre-cast concrete it easy to perform these steps. Pouring must be done in a shaded area.

FIGURE 7.24   Pouring raft foundation by using pump.

FIGURE 7.25   Pouring part of raft foundation.

FIGURE 7.26   Pouring concrete slab by pump.

In some cases a caisson can be used instead of a pump hose to pour the concrete. The caisson collects the concrete on the ground level and is lifted by the lifting tool, and the caisson gate is opened to pour the concrete, as shown in Figure 7.27.

7.6.1   POURING PUMPING CONCRETE

In modern countries pumping is the common method for pouring concrete especially in high-rise buildings. As shown in Figure 7.28, the concrete is poured in a raft foundation by using a pump.

The first concrete pump was used in America in 1913, and in 1930 many developed countries were using the pump for pouring by using a special valve design with a sliding door. From 1950 to 1960 pump concrete was used widely in Germany where about 40% of pouring was performed by the pumps.

Some companies in Germany competed to develop this pump. Schwing, Putzmeister, and Elba and others focus their development and research on valves.

The builders for the Jin Mao Building in Shanghai, China, boasted of pumping high strength concrete as high as 1200 ft (366 m). Pumping is limited by the plastic qualities of concrete, the type of pump available, and the piping needed to carry the product up to the desired level. For such great heights, a high-pressure unit is needed. Great thought must be given to the properties of concrete and how it will react when pressure is applied in a pipe. All these factors demanded innovations in concrete technology.

Pumping is a very efficient and reliable means of placing concrete, which makes it a very economical method as well. Sometimes, a pump is the only way of placing

FIGURE 7.27   Pouring beam by using caisson.

concrete in a certain location, such as a high-rise building or large slabs where the chutes of the concrete truck cannot reach where the concrete is needed.

The ease and speed of pumping concrete make it the most economical method of concrete placement. The Schwing KVM 55 can pump over 200 cubic yards per hour.

7.6.1.1   Incorrect Mix

The most common mix problem is concrete that does not retain its mixing water. Concrete can bleed due to poorly graded sand that allows water to bleed through the small channels formed due to voids in the sand or is too wet.

Insufficient mixing can cause segregation in the mix. For successful pumping, aggregate must have a full coating of cement grout to lubricate the mix as it is pumped.

A delay in placing the concrete due to traffic or job site problems, as well as hot weather conditions, may cause the concrete to set prematurely. This creates a mix that may be too stiff to pump because it will not fill the pumping cylinders, causing excessive pumping pressures.

7.6.1.2   Problems with Pipeline

The entire pumping system must be evaluated for the job. A properly sized system including pump capacity and motor horsepower to move the concrete through the full length of the pipeline is required.

Dirty pipe may cause blockages where old concrete has set, and may cause bleeding and segregation. Defective couplings, gaskets, or weld collars also can result in the loss of grout.

Look for bends that are too short, too sharp, or too numerous, all of which increase concrete pumping pressure. Variations of pipeline diameter, such as when a larger diameter hose is coupled with a smaller one, may cause blockages or rock jams because the concrete cannot flow as quickly through the smaller diameter pipeline.

FIGURE 7.28   Preparing concrete surface.

7.6.1.3   Operator Error

The most common error by inexperienced operators is setting up the pumping system improperly. Operators must know how to set up each job so that pipe or hose only needs to be removed, not added on. If the placing crew must add hose when a pour is in progress, the dry conditions inside the added hose are likely to cause a blockage.

Careless handling of flexible rubber discharge hoses can also be a problem, since kinking can occur. A rock jam is likely to be the result of a kinked hose, as the inside hose diameter is reduced, which restrains the aggregate in the line while the lubricating grout is allowed to pass. Premature localized wear of the hose, and eventual rupture of the hose, may also occur at the point where the hose is kinked.

7.6.2   Construction Joint

The locations of the construction joints should be known before and agreed on between all the concerned parties. The contractor will define the construction joint before work starts based on his equipment and manpower.

Generally, it is preferred to have no construction joints. Joint issues should be defined before the day of pouring and agreed on because a joint will be the weak point in the whole building.

The construction joints in beams and slabs are located at the point of zero bending moment or at the location of minimum shear force. The joint should be perpendicular to the effected internal forces, so the location of the construction joint should be in the workshop drawings. Also to be specified are the steel bars that will transfer tension and shear force in the joints and whether the designer will use dowels for joints.

On the second day, before pouring the concrete, the surface of the concrete will be cleaned by blowing air and any loose materials will be removed, and then water will be sprayed on the concrete surface. The old method is to use slurry, which is a mixing of water and cement, and paint the surface before pouring the concrete. The second preferable method is to use epoxy that bonds the new concrete to the old concrete.

7.7 COMPACTION PROCEDURE

Compaction is a very important factor as it enhances the concrete strength. During the compaction process make sure that there are no high vibrations to the concrete that has been poured, and do not change the form dimensions or the locations of the steel bars. The compaction must be performed by a competent person who uses the vibrator efficiently and perpendicular to the concrete as recommended by ACI as shown in Figures 7.29–7.31. Compaction should not stop until there are no more air bubbles.

Self-compacted concrete can be made using additives and silica fume, and this concrete will not need to undergo the compaction process. Self-compacted concrete was discussed in detail in Chapter 6. Note that self-compacted concrete is of more

FIGURE 7.29 Finishing concrete surface.

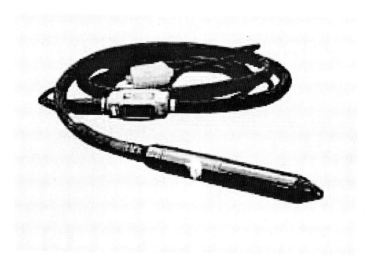

FIGURE 7.30 Concrete vibrator.

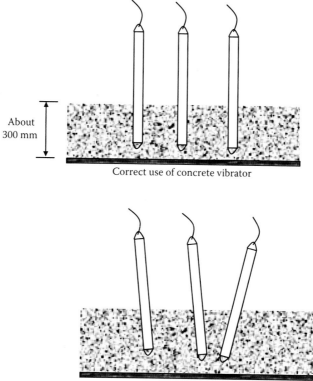

About
300 mm

Correct use of concrete vibrator

Incorrect use of the concrete vibrator

FIGURE 7.31 Correct and incorrect use of vibrator.

benefit due to the reduction of work on site, which enhances the concrete quality as a result.

7.8 EXECUTE CURING

Recently poured concrete must be protected from rain and fast drying as a result of warm or dry air or severe storms. The protection will be achieved by covering the cast concrete with a suitable coating system after finishing the pouring process until the time for final hardening, after which the curing process may proceed, which has already been defined by the designer and stated clearly in the project specification document.

Figures 7.32–7.34 illustrate the impacts of relative humidity and air temperature with wind speed on the speed of concrete drying.

The purpose of the curing process is to maintain wetness for not less than seven days in the case of ordinary Portland cement and not less than four days in the case of fast-hardening concrete or when using additives to accelerate the setting time. There are many ways of executing the curing process:

- Spraying water free from salt and any harmful substances.
- Covering the concrete surface with rough sand or wood manufacturer wastes and keeping it wet by spraying water on it regularly.
- Covering the concrete surface with high-density polyethylene sheets.
- Using additives that will be sprayed on the concrete surface.
- Using steam for curing in some special structures.

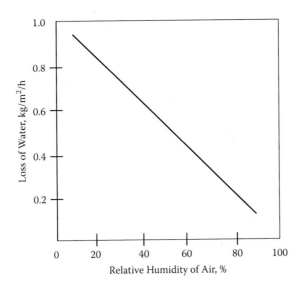

FIGURE 7.32 Relation between relative humidity and loss of water. *Source:* From Neville, A. M. 1975. *Properties of concrete.* London: Pitman.

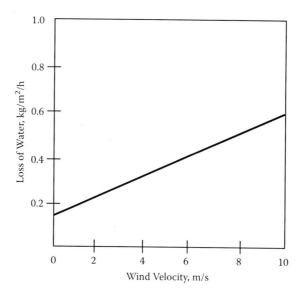

FIGURE 7.33   Relation between wind speed and loss of water. *Source:* From Neville, A. M. 1975. *Properties of concrete.* London: Pitman.

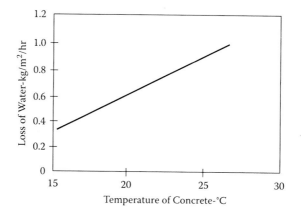

FIGURE 7.34   Relation between temperature degrees and loss of water.

FIGURE 7.35 Curing by using textile.

Figure 7.35 represents the second most common method of curing by covering the slab with a textile that is wet all the time.

The most common curing method is spraying water on the concrete member in the early morning and at night, and avoiding spraying water at sunrise as this will produce some cracks on the concrete surface due to evaporation of water.

The above method is used for normal environmental conditions, but not in some areas such as in the Middle East where the temperature in summer may reach to 55°C. In the construction industry a different technique is used which is to spray a chemical on the surface to prevent evaporation of the water from the concrete mix, as shown in Figure 7.36.

FIGURE 7.36 Spraying chemical.

FIGURE 7.37   Plastic sheet to protect from evaporation.

Another method is shown in Figure 7.37, which is covering the concrete member with a special plastic sheet to avoid the evaporation of water.

There are other methods, such as spraying waste from the wood manufacturing process and distributing it on the slab and keeping it wet all the time.

A similar idea is to distribute sand on the slab and spray it with water to be always wet, but the sand with water has a high dead weight that can affect the building at an early age.

The curing process costs little compared to the cost of concrete as a whole, but it increases the resistance of concrete in a very significant way and produces a tremendous increase in the concrete strength.

Figure 7.38 shows that the concrete strength clearly increases when the curing process takes 7 days instead of 3 days curing, and the effect will be greater when the duration of curing is 14 days. This increase in strength of concrete will be continuous throughout the structure's life.

The increase of concrete strength due to curing after 28 days has no more effect than curing for only 14 days. The specifications of the project should clearly define exactly the required curing time as it differs from project to project according to the weather factors of the area in which the project is located and according to the concrete members.

7.8.1   Curing Process in ACI

According to ACI 5.11, the curing of concrete shall be maintained above 50°F and in a moist condition for at least the first 7 days after placement; in the case of high-early-strength concrete shall be maintained above 50°F and in a moist condition for at least the first 3 days.

If there is a need to accelerate curing, high pressure steam, steam at atmospheric pressure, heat and moisture, or other accepted processes shall be permitted to accelerate strength gain and reduce time of curing.

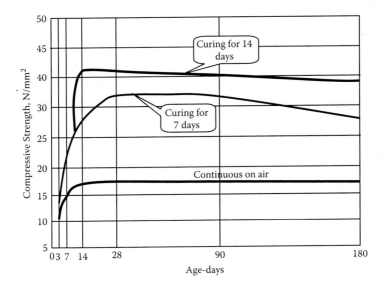

FIGURE 7.38 Relation between curing time and compressive strength.

Accelerated curing will provide a compressive strength of the concrete at the load stage considered at least equal to the required design strength at that load stage.

The curing process shall produce concrete with durability at least equivalent to that required by the engineer or architect; strength tests shall be performed to ensure that curing is satisfactory.

The compressive strength of steam-cured concrete is not as high as that of similar concrete continuously cured under moist conditions at moderate temperatures.

Also the elastic modulus Ec of steam-cured specimens may vary from that of specimens moist-cured at normal temperatures. When steam curing is used, it is advisable to base the concrete mixture proportions on steam-cured test cylinders.

Accelerated curing procedures require careful attention to obtain uniform and satisfactory results. Preventing moisture loss during the curing is essential.

In addition to requiring a minimum curing temperature and time for normal- and high-early-strength concrete, the adequacy of field curing must be determined. At the test age for which the strength is specified (usually 28 days), field-cured cylinders should produce strength not less than 85% of that of the standard, laboratory-cured cylinders. For a reasonably valid comparison, field-cured cylinders and companion laboratory-cured cylinders should come from the same sample. Field-cured cylinders should be cured under conditions identical to those of the structure. If the structure is protected from the elements, the cylinder should be protected.

7.8.2 Curing in British Standard

Based on BS, curing prevents the loss of moisture from the concrete while maintaining a satisfactory temperature regime. The curing protocol should prevent the development of high temperature gradients within the concrete.

TABLE 7.15

Minimum Periods of Curing and Protection (BS8110)

		Min. Period of Curing (Days)	
Type of Cement	Condition after Casting	5°C to 10°C	Any Temperature between 10°C and 25°C
Portland cement and sulfate-resisting Portland cement	Average	4	$60/(t + 10)$
	Poor	6	$80/(t + 10)$
All cement except the above and super sulfated cement	Average	6	$80/(t + 10)$
	Poor	10	$140/(t + 10)$
All	Good	No special requirements	

Notes: Good, damp and protected (relative humidity greater than 80%; protected from sun and wind); average, intermediate between good; poor, dry or unprotected (relative humidity less than 50%; not protected from sun and wind).

The rate of strength development at early ages of concrete made with super sulfated cement is significantly reduced at lower temperatures. Super sulfated cement concrete is seriously affected by inadequate curing and the surface has to be kept moist for at least 4 days. Curing and protection should start immediately after the compaction of the concrete to protect it from the following:

- Premature drying out, particularly by solar radiation and wind
- Leaching out by rain and flowing water
- Rapid cooling during the first few days after placing
- High internal thermal gradients
- Low temperature or frost
- Vibration and impact, which may interfere with bonding of concrete to the reinforcement

Where members are of considerable bulk or length, the cement content of the concrete is high, the surface finish is critical, or special or accelerated curing methods are to be applied, the method of curing should be specified in detail.

BS 1881 states that the surfaces should normally be cured for not less than times shown in Table 7.15. Depending on the type of cement, the ambient conditions, and the temperature of the concrete, the appropriate period is taken from Table 7.15 or calculated from the last column of that table. During this period, no part of the surface should fall below a temperature of 5°C.

The surface temperature depends upon several factors, including the size and shape of the section, the cement class and cement content of the concrete, the insulation provided by the formwork or other covering, the temperature of the concrete at the time of placing, and the temperature and movement of the surrounding air. If not measured or calculated, the surface temperature should be assumed equal to the temperature of the surrounding air (see CIRIA Report No. 43). The most common methods of curing as specified under British specifications are as follows:

- Maintaining formwork in place
- Covering the surface with an impermeable material such as polyethylene, which should be well sealed and fastened
- Spraying the surface with an efficient curing membrane

7.8.3 PROTECT SPECIAL STRUCTURES

The code and specifications requirements in the design and execution of concrete structures to protect the structure from corrosion are not sufficient in structures exposed to atmospheric factors that cause corrosion, such as in offshore structures and parts of structures exposed to moving water (splash zone), or in parking garages or bridges that require salt to melt ice.

Modern commercial or residential buildings adjacent to seawater may not be exposed to seawater directly but are exposed to chlorides in marine areas with a warm atmosphere, such as in the Middle East, especially the Gulf area.

There are high investments in structures on islands or quasi islands, which are surrounded by water from all sides in addition to high temperature, as in Singapore and Hong Kong, where corrosion starts early and the corrosion rate is very high also.

Therefore, the control of corrosion through the use of good concrete, appropriate content of the cement and the specific w/c ratio, and caution in the process of curing and during construction are key factors for the protection of reinforced concrete structures from corrosion. They form the first line of defense to attack corrosion.

REFERENCES

ACI Committee 201. 1994. Guide to durable concrete. In *Manual of concrete practice, part 1*. Detroit, MI: American Concrete Institute.

ACI Committee 301. 1994. Specification for structural concrete for building. In *Manual of concrete practice, part 3*. Detroit, MI: American Concrete Institute.

ACI Committee 357. 1994. Guide for design and construction of fixed off-shore concrete. Detroit, MI: American Concrete Institute.

Arnon, B., S. Diamond, and N. S. Berke. 1997. *Steel corrosion in concrete*. London: E&FN SPON.

Beeby, A. W. 1979. Concrete in the oceans-cracking and corrosion. Technical report no. 2 United Kingdom: CIRIA/UEG, Cement and Concrete Association.

British Standard 8110:1985. Structural use of concrete, part 1. Code of practice for design and construction.

British Standard PD 6534:1993. Guide to use in the UK of DD ENV 206:1992 "Concrete: Performance, Production, Placing and Compliance Criteria."

Browne, R. D., M. P. Geoghegan, and A. F. Baker. 1993. Analysis of structural condition from durability results. In *Corrosion of reinforcement in concrete construction*, A. P. Crane, ed., 193–222. United Kingdom: Society of Chemical Industry.

Building code requirements for reinforced concrete (ACI 318-89) and commentary (ACI 318R-89). 1994. In Manual of concrete practice, part 3. Detroit, MI: American Concrete Institute.

CIRIA, Technical Note 43. Construction Industry Research and Information Association (CIRIA), London, 1987.

Egyptian Code of Practice, 2003.

El-Reedy, M.A. and Mohamadine. H. 2005. *Concrete Quality Control Related to Management Performance,* Montreal Conference, CSCS.

European pre-standard ENV 206. 1992. Concrete-performance. Production, placing and compliance criteria.

Neville, A. M. 1975. *Properties of concrete.* London: Pitman.

Van Daveer,J.R. 1975. *Techniques for evaluating reinforced concrete bridge decks,* ACI Journal,72(12), 697–704.

8 Structure Integrity Management

8.1 INTRODUCTION

After a long period of reinforced concrete being used, some defects and problems appeared in various concrete structures worldwide, repudating the idea that the main advantage of using concrete is that it will not need maintenance because it is durable material. The presence of defects after various periods of time depends on the structure type and the surrounding environmental conditions.

Old structures were left for long periods without maintenance until large cracks were seen, along with partial collapse of the concrete cover, for example.

This inaction significantly increased the cost of repair to restore the initial structure's capacity to carry loads safely.

A very clear example of the high cost is seen in repair of bridges.

This led to the development of integrated plans for the maintenance of concrete structures in order to achieve the highest return or be less expensive.

8.2 STRUCTURE INTEGRITY MANAGEMENT

Structure integrity management is now used worldwide to maintain structures in a safe condition along their lifetime. The system requires a fund, competent persons, and tools such as new software to collect the data, perform analysis, and track the inspection and maintenance plan.

In general, this system requires one team or more with specialists in different areas based on the structure type. The structure integrity delivery process is summarized in the following generic steps:

- Developing plan and schedules
- Defining monitoring requirements
- Implementing the monitoring scheme
- Technical assessment
- Recommending change
- Updating plans and schedules

Figure 8.1 presents the structure integrity management system cycle. The core must be the organization goals and target, and if a third party is going to establish the system it must know clearly the goals for the short term and the long term,

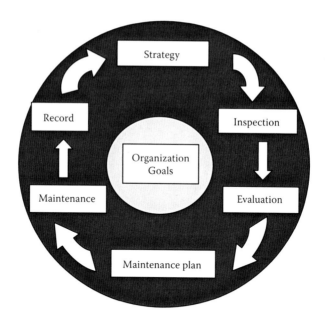

FIGURE 8.1 Structure integrity management system cycle.

for example, the structure's lifetime and its age now and time during which the structure integrity system will be developed.

First of all, develop the strategy to perform all the activities while looking at the organization's goals. The second step is to perform the inspection specified in the strategy plan. Then go through the evaluation of the structure, which may be performed by using restructure analysis in a complicated project or evaluation by a competent engineer.

The maintenance plan will be developed based on the evaluation results and the available budget. Then go through the schedule of major repair or minor repair. Finally, record the status of the structure's condition and revise the inspection strategy.

Abd El-Kader and Al-Kulaib presented a case study in 1998. This system was developed in most countries for integrated infrastructure management systems especially for bridges, pavement, sanitary sewer networks, and storm sewer networks.

For example, the POTIS bridge management system software, developed by the Federal Highways Administration (FHWA) in association with the American Association of State Highways and Transportation Officials (AASHTO), was selected. This is a systematic software tool that enables staff to plan, organize, and track bridge information and maintenance and construction needs, plan projects, and allocate budgets. This includes inventory data, inspection data, condition rating, deterioration rates, load ratings, and structural and functional appraisals.

POTIS provides a powerful way to predict deterioration; identify bridge maintenance, replacement, and rehabilitation needs; develop and compare maintenance replacement and rehabilitation strategies for bridges and bridge systems; and allocate and track funds for maintenance replacement and rehabilitation programs. It allows comparison of the funds and budget and the maintenance or rehabilitation plan.

For this management system for bridges or software that deals with different types of structures, the following items must be input:

- Structure identification (number, location, type, and name)
- Rating (according to local road condition in case of bridges)
- Environmental conditions
- Importance of the structure
- Historical data
- Structure condition

8.3   RISK-BASED MAINTENANCE STRATEGY

In general, the economic factor is one of the most influential. The cost of repairing the concrete structure, as well as the selection of the suitable protection system to protect the steel reinforcement from corrosion, are the major factors that control the ways to choose repair or protection methods.

Virtually every method of repair and protection has an expected lifetime. By knowing the whole structure lifetime, it can be easy to calculate the number of times expected maintenance must be done through the structure's lifetime. The economic study to choose alternatives in the case of a new structure is very important to determine the appropriate method for protection by taking into account the lifetime of the protection method with respect to the whole life of the structure along with the initial costs of periodic maintenance.

Generally, the cost calculation is based on a summation of the initial costs of protection and the cost of maintenance that will be performed in different periods and periodically. The number of maintenance times during the whole life of the structure varies, depending on the method of protection as well as the method used in the repair procedure during maintenance.

To maintain an existing old building and restore the structure to its original strength, the repair method of rehabilitation is also governed by the initial costs of repair, as well as the number of times the maintenance will be done and its costs over the remaining life of the structure.

El-Reedy and Ahmed (1998) explain how to choose among the various alternatives for the protection methods as well as the appropriate repair methods and the materials that are usually used in repairs. The previous discussion is from the technical point of view only; however, in this chapter, the way to compare alternatives is discussed from an economic view to assist in the decision-making procedure.

Therefore, the appropriate method will study the economics and compare alternatives to clarify the economic factors that affect the method of calculation. The lifetime of maintenance or the protection of the structure must be taken into consideration, so the required time to perform the maintenance must be discussed thoroughly, and the first step is to calculate this time.

This will also be applied to a practical example to illustrate the comparison of alternatives. The cost difference between the methods of protection for different structures will be illustrated.

Determining the time of maintenance depends on the maintenance cost estimate versus the structure probability of failure; therefore, it is important to select the appropriate time to perform maintenance.

The method of making this decision from the standpoint of determining the right time for the maintenance, which is a less expensive verification process, is called optimization procedure.

8.3.1 COST CALCULATION BASIC RULES

There are several basic rules for the calculation of the economic costs of any engineering project, as well as applying those rules to select the type of protection required of a structure and method of repair.

The most popular methods of economic analysis tools are the present value, future value, and the interest rate of return and other methods. Here, the method of calculating the present value is briefly described, as it is the easiest way to select the appropriate repair, as well as selecting the appropriate system to protect the structure from the corrosion effect.

8.3.2 PRESENT VALUE METHOD

The cost of protecting reinforced concrete from corrosion consists of the preliminary costs of protection paid at the beginning of construction; on the other side, the cost of maintenance and repair will be over the lifetime of the structure.

In many cases the cumulative cost of maintenance and repair is higher than initial costs. In many projects cost calculation is often based on the initial costs only, but the result is that the total cost is very high compared to the initial cost estimate.

This method is used to calculate the present value of future repair, including the cost of equivalent current value with the assumption that the repair will take place after a number of years.

$$\text{Present value} = \text{repair cost} \, (1 + m)^{-n} \tag{8.1}$$

where m is the discount rate, which is the interest rate before inflation rate, for example, assuming that the interest rate is 10% and the inflation rate of 6%, the discount rate (m) is equal to 4% or 0.04, and n is the number of years.

The whole structure cost consists of the initial cost, called capital cost (CAPEX), in addition to the sum of the present values of future costs due to maintenance, called operating cost (OPEX). When there is an increase in the rate of inflation, the cost of future repair has no effect. But if the inflation rate decreases it will increase the present value of future repair.

In this chapter, it is assumed that the value of the inflation rate is about 4%. The inflation rate depends on the country's general economics, and every country has its own inflation rate, which is published.

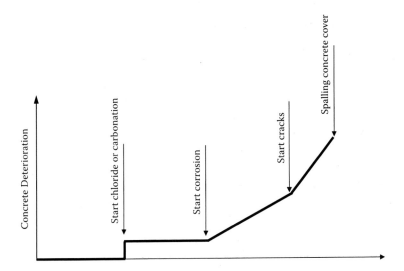

FIGURE 8.2   Concrete structure deterioration process.

8.3.3   REPAIR TIME

The time required to repair the structure as a result of steel corrosion is the time at which the beginning of corrosion in the steel reinforcement bars takes place in addition to the time needed to spall the concrete cover, which is a sign of deterioration that essentially requires repair.

Tutti in 1982 explained corrosion with time. We can note that these steps apply to all types of corrosion, and there is no difference for chloride attack or carbonation propagation, but these processes take a long time to break the passive protection layer on the steel bars and start corrosion.

After that, additional time will be added from the beginning of corrosion to a significant deterioration in the concrete, for which necessary repair will be required. Raupach's research in 1996 pointed out that in the case of concrete bridges, degradation occurs in about 2 to 5 years. Therefore, the time for repair is the total time required for the protection of depassivation in addition to 3 years.

The steps of corrosion's effect on concrete structures are illustrated in Figure 8.2. Chloride concentration or carbonation will be accumulated on the surface and then propagated into the concrete, as shown in the figure, as a start of chloride or carbonation. The next step is the propagation of chlorides or carbonation until they reach the steel bars, and the third step is the start of corrosion on the steel bars, which will have an impact on the concrete strength by reduction of steel diameter and cracks on the concrete surface. The last step is an increase in the crack width until spalling the concrete cover.

The time required for repair depends on the time it takes to increase the percentage of chloride concentration to the limit that will initiate the corrosion in addition to the rate of corrosion.

In the previous chapter several types of protection against corrosion were studied. These different methods will extend the start of corrosion as they will reduce

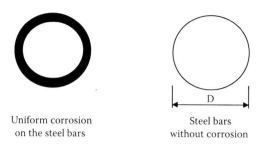

Uniform corrosion
on the steel bars

Steel bars
without corrosion

FIGURE 8.3   Reduction in steel diameter due to uniform corrosion.

the rate of chloride or carbonation propagation in concrete and reduce the rate of corrosion after that. Note that the above analysis relies on the non-interference of the hair cracks on the concrete in the rates of propagation of chlorides or carbonation within the concrete, where it is assumed that the design was based on the absence of an increase in the cracks exceeding that permissible in codes, as well as assuming the concrete is executed based on the quality control procedure according to code and thus assuming cracks within the allowable limits.

The time required to start the repair depends on the surrounding weather and environmental factors that affect the beginning rate of corrosion.

This time is determined by knowing the rate of corrosion and the required time to spall of the concrete cover. The deterioration of concrete increases the probability of the collapse of the structure with time, and some studies have identified the probability of failure, which should not be beyond structure reliability, which is classified in various specifications.

A study by El-Reedy et al. in 2000 on residential buildings focused on determining the appropriate time to perform repair due to corrosion on concrete columns, taking into account environmental conditions around the structure, which are affected by humidity and temperature and impact rates of corrosion.

A corrosion rate of 0.064 mm per year reflects the dry air while the rate of corrosion of 0.114 mm per year is based on high moisture, taking into account the increasing resistance of the concrete with time as well as the method of determination for high steel or low steel columns with different times required for the repair process.

8.3.4   Capacity Loss in Reinforced Concrete Sections

The fundamental of design of any reinforced concrete member is that the capacity of the reinforced concrete member is dependent on the cross section dimensions (concrete and steel area) and material strength (concrete strength and steel yield strength). In the case of uniform corrosion, as shown in Figure 8.3, the total longitudinal reinforcement area can be expressed as a function of time, t, as follows:

$$As(t) = \begin{cases} n\pi D^2 / 4 & \text{for } t \leq T_i \\ n\pi \left[D - 2C_r(t - T_i) \right]^2 / 4 & \text{for } t \leq T_i \end{cases} \qquad (8.2)$$

where D is the diameter of the bar, n is number of bars, T_i is time of corrosion initiation, and C_r = rate of corrosion. Equation (8.2) takes into account the uniform corrosion propagation process from all sides.

In Figures 8.4 and 8.5, note that over time, the collapse of the structure is more likely with the increase in the rate of corrosion.

Reinforced concrete columns that are concentrically loaded will be due for maintenance in 4 to 5 years from the initial time of corrosion based on the lowest steel ratio and higher corrosion rate. In cases of higher steel ratios and lower corrosion rates, this period may increase to 15 to 20 years.

Moreover, this study predicts that a moment on a column with eccentricity will increase, so we increase the percentage of steel. In this case, it is expected that the structure will deteriorate and move toward criticality after a couple of years in the case of lowest reinforcing steel ratio and very high corrosion rate; on the other hand,

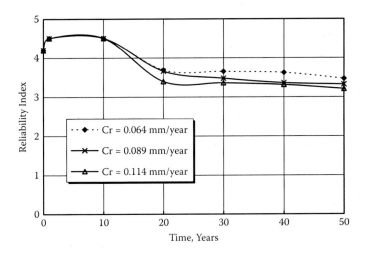

FIGURE 8.4   Effects of corrosion rate on reliability index at reinforcement ratio equal to 1%.

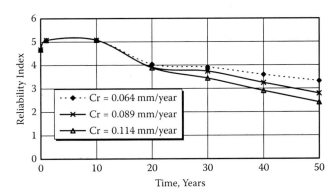

FIGURE 8.5   Effects of corrosion rate on reliability index at reinforcement ratio equal to 4%.

in the case of higher reinforcing steel ratio and lower corrosion rate, the maintenance will be due after about 6 years.

As expected, the required time for the repair work is closely linked to the nature of the structure and the method of design, as well as the importance of the building. Vital structures such as nuclear facilities need ways to protect them along their lifetime much differently from other structures such as residential buildings.

We will apply the method of calculating costs through the example of protecting a reinforced concrete foundation in a petrochemical process plant near the Red Sea, taking into account the same protection methods as used by Arnon et al. (1997) when explaining the economic study in a surface bridge exhibition to corrosion where salt is always used to melt ice. So the probability of attack by chlorides is very high. The cost of repair and protection methods in the example is roughly based on the cost in Egypt. When applied in different countries, it will vary, but the price presents a comparison of different ways of protection in terms of cost.

8.4   EXAMPLE

The example is a concrete foundation in a petrochemical process plant near the sea with a concrete cover of 50 mm for all foundations. The life span of the plant is 75 years, and ready mix concrete was used with a water/cement ratio (w/c) equal to 0.4. The average daily temperature is about 26°C. Taking into account the various types of protection used, including painting steel with epoxies or the use of silica fume or corrosion inhibitor, the initial cost estimate for the protection using those methods is shown in Table 8.1 and the value of cost is calculated at the rate of the American dollar per cubic meter.

The cost of repairs is assumed to be about $200 per m^3. Assume that the traditional repair method is used, which is to remove the cover and apply the normal repair procedure.

From Table 8.1, the corrosion inhibitor type (1) uses an anodic type such as calcium nitrate with concentration 10 kg/m^3, and (2) calcium nitrates with concentration 15 kg/m^3. The values of chloride ion content on the steel level are shown in Table 8.1. Increasing the corrosion inhibitor concentration will increase the time period to start corrosion.

Note that the discount rate is imposed, as we mentioned earlier, at about 4% with the assumption that repair will be sufficient for 20 years only.

8.4.1   Required Time to Corrosion

In structures exposed to chlorides as in the previous example, the time required for corrosion is the time required for the spread of chlorides in the concrete until they reach the steel. At a certain chloride level, corrosion will start.

The assumed rate of diffusion is about 1.63×10^{-12} m^2/sec for concrete that was cast using the ratio of w/c of about 0.4. It is assumed that the propagation rate is about 1.03×10^{-12} m^2/sec for silica fume. The propagation rate and its impact are shown in Figure 8.2.

TABLE 8.1
Comparison of Ways to Protect Steel

Method of Protection	Cost ($/m³)
No protection	—
Corrosion inhibitor (1)	40
Corrosion inhibitor (2)	50
Silica fume	80
Epoxy coated steel bars	55
Silica fume + corrosion inhibitor (1)	120
Silica fume + corrosion inhibitor (2)	130
Epoxy coated steel bars + corrosion inhibitor (1)	95
Epoxy coated steel bars + corrosion inhibitor (2)	105
Cathodic protection system	250

TABLE 8.2
Time Required to Start Corrosion

Method of Protection	Chloride Limit (kg/m³)	Time to Reach Limit	First Repair Time	Second Repair Time	Third Repair Time
No protection	0.9	17	20	40	60
Corrosion inhibitor (1)	3.6	37	40	60	
Corrosion inhibitor (2)	5.9	+75			
Silica fume	0.9	22	25	45	65
Epoxy coated steel bars	0.9	17	32	52	
Silica fume + corrosion inhibitor (1)	3.6	51	54		
Silica fume + corrosion inhibitor (2)	5.9	+75			
Epoxy coated steel bars + corrosion inhibitor (1)	3.6	37	52		
Epoxy coated steel bars + corrosion inhibitor (2)	5.9	+75			

Time is needed to increase the content of the chlorides on the steel bars as a result of the previous propagation of chloride inside the concrete cover whose thickness is 50 mm.

Calculations showing the beginning of corrosion through computational methods based on Berke et al. in 1996 are shown in Table 8.2. The time needed to reach that amount of chlorides then begin corrosion differs from one protection system to another. Note that there is no increase in the limits of chlorides in the corrosion of steel coated by epoxies or by using silica fume.

There are several equations to calculate the time required for the propagation of carbonation inside concrete cover in Table 8.3. To calculate the depth of the transformation of carbon use the following equations:

TABLE 8.3

Cost Analysis for Various Protection Methods ($/m³)

Protection Method	Initial Cost	First Repair (NPV)	Second Repair (NPV)	Third Repair (NPV)	Total Repair Cost	Total Cost
No protection	0	91.3	41.65	19.03	151.98	151.98
Corrosion inhibitor (1)	40	41.66	19.03		60.69	64.69
Corrosion inhibitor (2)	50	0				50
Silica fume	80	75	42.22		117.22	197.22
Epoxy coated steel bars	55	57.03	26		83.03	138.03
Silica fume + corrosion inhibitor (1)	120	24.05			24.05	144.05
Silica fume + corrosion inhibitor (2)	133	0			0	133
Epoxy coated steel bars + corrosion inhibitor (1)	95	26			26	121
Epoxy coated steel bars + corrosion inhibitor (2)	105	0				105

$$d = A(B)^{-0.5} \tag{8.3}$$

where A is a fixed amount depending on the permeability of concrete, as well as the quantity of carbon dioxide in the atmosphere and several other factors.

$$A = (17.04(w/c) - 6.52) \cdot S \cdot W \tag{8.4}$$

where w/c is the ratio of water to cement (less than 0.6), S is the effect of cement type, and W is the weather effect. These equations give the average depth of the transformation of carbon. Therefore, when calculating the maximum depth of the transformation of carbon it should be increased from 5 to 10 mm. S = 1.2, in the case of the use of cement by 60%. W = 0.7 for concrete protected from the outside environment.

8.4.2 Time Required to Deteriorate

The time required to start corrosion has already been studied. This time is about 3 years in the absence of corrosion inhibitor, and it extends up to 4 years in the case of adding corrosion inhibitor. On the other hand, in the case of the use reinforcing steel coated by epoxy the period is extended to 15 years.

The use of reinforcing steel coated by epoxy helps reduce the rate of corrosion in a clear reversal of contraceptive use of corrosion inhibitor preventive.

Generally, the time of the collapse of the concrete cover after the beginning of corrosion depends on the rate of corrosion in the steel reinforcement. The rate of corrosion is closely related to relative humidity, as stated by Tutti in 1982, who also stated that in steel corrosion the rate of carbonation results in less corrosion, at relative humidity of 75% and less, and that the rate of corrosion increases very quickly upon the arrival of relative humidity to 95%. When temperatures drop below the rate of corrosion, in temperature about 10 degrees below, the rate of corrosion decreases by about 5% to 10%.

Broomfield (1997) stated that the rate of corrosion is affected by the relative moisture and the proportion of chlorides. As stated by Moringa (1998), corrosion is totally prevented at relative humidity less than 45%, regardless of the content of chlorides and the temperature or oxygen concentration.

Broomfield also stated that cracks occur when steel measures about 0.1 mm and has been seen at smaller dimensions as well. Cracks depend on the oxygen concentration and distribution as well as the ability of concrete to withstand excess stresses.

Shortage in a sector of 10 or 30 μm results in enough corrosion of the fragile layer to form a chasm in the concrete.

The following equation by Kamal et al. (1992) is used to calculate the time necessary for the emergence of the effects of corrosion:

$$t_s = 0.08(C-5)/(D \cdot C_r) \tag{8.5}$$

where t_s is the time of the beginning of corrosion until the fall of the concrete cover, C represents the concrete cover thickness, D is the steel bar diameter, and C_r is the corrosion rate in mm/year.

The total time expected to perform the first repair depends on the time of the beginning of corrosion in addition to the time needed to increase the corrosion to produce deterioration of the concrete time to perform the repair.

8.4.3 Cost Analysis for Different Protection Methods

The equation for the calculation of the present value of all methods of protection for steel reinforcement is outlined in Table 8.2. The different present values of methods of protection were clarified in Table 8.3, and initial costs and the cost of repair have also been clarified. The total cost calculations of the current value are shown in Figure 8.6 for each method of protection.

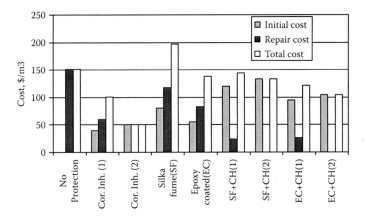

FIGURE 8.6 Economic comparison of different alternatives.

TABLE 8.4

Comparison of Initial Cost, Total Repair Cost, and Total Cost for Various Methods

Protection Method	Initial Cost	Total Repair Cost	Total Cost
No protection	0	151.98	151.98
Corrosion inhibitor (1)	40	60.69	100.69
Corrosion inhibitor (2)	50		50
Silica fume	80	117.22	197.22
Epoxy coated steel bars	55	83.03	138.03
Silica fume + corrosion inhibitor (1)	120	24.05	144.05
Silica fume + corrosion inhibitor (2)	133	0	133
Epoxy coated steel bars + corrosion inhibitor (1)	95	26	121
Epoxy coated steel bars + corrosion inhibitor (2)	105		105

The cost of cathodic protection is about \$500 to \$1500/m^3. Taking into account the lower cost of \$500/m^2 it will be much higher than the cost of other alternatives as shown in Table 8.4.

If the cathodic protection system is applied after 20 years it would have an NPV equal to around \$450 based on \$1000/m^3 cost and no operating, maintenance, or repair costs for the next 55 years.

A comparison of Tables 8.3 and 8.4 can help you choose the least costly method of protection. Any comparison must take into account the assumptions on which the calculations are imposed, including inflation and interest rates as well as the prices of labor and raw materials.

8.5   REPAIR AND INSPECTION STRATEGY AND OPTIMIZATION

The decision-making methods have become the focus of numerous studies. The main goal is to focus on the economic factors representing cost of the project as a whole, because the wrong decision could cost a huge amount of money to reverse.

Therefore, some studies use a decision tree to determine the appropriate time for repair, particularly for bridges that need periodic inspections.

When devising a maintenance plan strategy based on cost we must consider all the factors that affect corrosion including the carbonation, chloride effect both internal and external, concrete cover, and quality. The target is to make a decision defining the time for regular maintenance, whether it will be every 10 years or 15 years or 20 years and so on. This decision is the first, and then the least expensive option is identified while looking at the operation philosophy at the same time.

Figure 8.7 summarizes the reliability of any structure during its lifetime. Early in its life a structure will be at its highest capacity and the probability of failure value is based on the code and standard by which the structure is designed.

Every structure will deteriorate, so its probability of failure increases, and after time Δt when inspection and repair have been performed the structure will recover

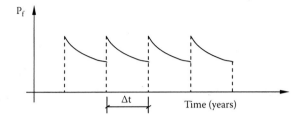

FIGURE 8.7   Concrete structure performance.

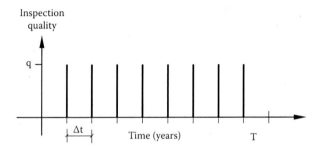

FIGURE 8.8   Inspection strategy.

its original strength, as shown in Figure 8.7, until the probability of failure increases to a certain limit and then another maintenance is performed and so on.

Inspection alone does not improve reliability unless it is accompanied by a corrective action for a discovered defect. Some strategies used in a wide range of concrete structures include

- Repair after monitoring until the crack depth reaches a certain proportion of the material thickness
- Immediate repair on detection of indications
- Repair at a fixed time (e.g., 1 year) after detection of indications
- Repair as new (i.e., welding)

Generally, it is assumed that inspections are performed at constant intervals, as shown in Figure 8.8, since the inspection authorities often prefer a constant inspection interval to facilitate the planning. Inspection intervals are chosen so that the expected cost of inspection, repair, and failure is minimized.

While most concrete structure inspection techniques are visual or involve NDT inspection methods, the ability to detect damage is dependent on the quality of inspection performed. A high quality inspection method will provide a more dependable assessment of damage. No repair will be made unless the damage is detected.

8.5.1   REPAIR

Inspection may not affect the probability of failure of the structure. Following an inspection, a decision must be made regarding repair if damage is found. The repair decision will depend on the inspection quality. With advanced inspection methods, even a small defect can be detected and repaired.

A high quality of inspection may lead to a higher quality of repair, which brings the reliability of the structure closer to its original condition. Aging has an effect and the reliability of the structure is decreased. We propose that after inspection and repair the structure capacity will be the same as the design conditions, as shown in Figure 8.5.

8.5.2   EXPECTED TOTAL COST

As mentioned in El-Reedy and Ahmed (1998), the first step is to determine the service life of the structure. Assume it is 75 years and routine maintenance is scheduled every 2 years (it starts at $t = 2$ years and is continuous until $t = 74$ years). Consequently, preventive maintenance work will be performed 37 times during the life of the structure. Therefore, the lifetime routine maintenance cost becomes

$$C_{FM} = C_{m2} + C_{m4} + C_{m6} + \dots + C_{m74} \tag{8.6}$$

where C_{FM} indicates the total maintenance cost, and the total expected cost in its lifetime (T) is based on the present worth value. The expected lifetime preventive maintenance cost becomes

$$C_{IR} = C_{IR2} \frac{1}{(1+r)^2} + C_{IR4} \frac{1}{(1+r)^4} + C_{IR6} \frac{1}{(1+r)^6} + \dots + C_{IR74} \frac{1}{(1+r)^{74}} \tag{8.7}$$

where
   C_{IR} = the periodic inspection and minor repair.
   r = net discount rate of money.

In general, for a strategy involving m lifetime inspections, the total expected inspection cost is

$$C_{ins} = \sum_{i=1}^{m} C_{ins} + C_R \frac{1}{(1+r)^{TI}} \tag{8.8}$$

where
   C_{ins} = inspection cost based on inspection method.
   C_R = repair cost.
   R = net discount rate.

Finally, the expected total cost C_{ET} is the sum of its components including the initial cost of the structure, the expected cost of routine maintenance, the expected cost of preventive maintenance, which includes the cost of the inspection and maintenance, and the expected cost of failure. Accordingly, C_{ET} can be expressed as

$$C_{ET} = C_T + (C_{ins} + C_R)(1 - P_f) + C_f \cdot P_f \tag{8.9}$$

The objective remains to develop a strategy that minimizes C_{ET} while keeping the lifetime reliability of the structure above a minimum allowable value.

8.5.3 OPTIMIZATION STRATEGY

To implement an optimum lifetime strategy, the following problem must be solved:

$$\text{Minimize } C_{ET} \text{ subjected to } P_{f\,life} \leq P_{max}$$

where P_{max} = maximum acceptable lifetime failure probability. Alternatively, considering the reliability index

$$\beta = \phi^{-1}(1 - P) \tag{8.10}$$

where ϕ is the standard normal distribution function, the optimum lifetime strategy is defined as the solution of the following problem:

$$\text{Minimize } C_{ET} \text{ subjected to } \beta_{life} \geq \beta_{min}$$

The optimal inspection strategy with regard to costs is determined by formulating an optimization problem.

The objective function (C_{ET}) in this formulation is defined as including the periodic inspection and minor joint repair cost, and the failure cost, which includes the cost of major joint repair. The inspection periodic time (Δt) is the optimization variable, which is constrained by the minimum index β specified by the code and the maximum periodic time. The optimization problem may be mathematically written as

Find Δt, which minimizes the objective function

$$C_{ET}(\Delta t) = (C_{IR})(1 - P_f(\Delta t)) \left(\frac{(1+r)^T - 1}{\left((1+r)^{\Delta T} - 1\right)(1+r)^T} \right)$$

$$+ C_f P_f(\Delta t) \left(\frac{(1+r)^T - 1}{\left((1+r)^{\Delta T} - 1\right)^T (1+r)^T} \right) \tag{8.11}$$

Subject to $\beta(t) \geq \beta^{min}$, $\Delta t \leq T$

where C_{IR} is the periodic inspection and minor repair cost per inspection; C_f is the major repair cost; i is the real interest rate, and β^{min} is the minimum acceptable reliability index. C_{IR} and C_f are assumed constant with time.

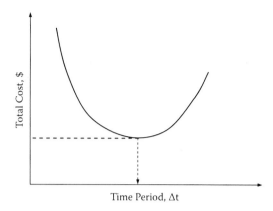

FIGURE 8.9   Optimization curve to obtain optimum maintenance period.

Even though the failure cost is minimized (as part of the total cost), it is often necessary to put a constraint on the reliability index to fulfill code requirements.

The time period for the proposed repair is T, n is the total lifetime of the building, and C_{IR} is the cost of inspection and repairs resulting from the collapse.

Moreover, the value of the expected cost at each period will be a curve that follows Figure 8.9, which defines the time required for periodic maintenance, which when achieved is less expensive, as shown in the curve.

Δt means the period required for regular maintenance that achieved the highest cost. The previous equation is generally used for comparison of different types of repair with different time periods and different cost.

The decision can be determined by the expected cost and risk of collapse of the concrete member or by the use of the approximate method as shown in Table 8.5. For the possibilities imposed at the structure's beginning, the probability of collapse is zero. The structure life time is assumed to be 60 years. The possibility of the building's collapse is 100% at 60 years. The probability of collapse at every period is shown in Table 8.5.

There are also some special software programs for the management of bridges (BMS) as identified by Abd El-Kader and Al-Kulaib in 1998. They define the time required to conduct inspection and maintenance, taking into account the rate of deterioration. There is a limitation on the time required for periodic inspection and maintenance work as well as to determine the cost, which depends on the budget planning for repairs. The program was developed by the FHWA.

Note that the maintenance plan and its implementation depend on the importance of the building. When proceeding with maintenance take into consideration the sequences of failure.

For example, there is a difference between the concrete columns and concrete slab around the building. Taking into account the repair cost of the two members, it may be the same or in some situations the repair of the concrete slab on the grade has a higher cost than repairing the columns, so what will you repair first? Start with repairing the concrete columns because of the danger that their failure will cause from an economic point of view.

TABLE 8.5
Relation of Structure Lifetime and Probability
of Failure

Life (years)	0	10	20	30	40	50	60
Probability of failure (%)	0	9	25	50	75	91	100

Regular maintenance within a certain interval (preventive maintenance) is important because it extends a structure's lifetime. Maintaining a reliable structure means
economic benefit, particularly in developed countries. Coastal cities represent huge
investments and are highly vulnerable to corrosion from chloride effects. They
require special maintenance plans.

In large cities, the amount of carbon dioxide in the atmosphere is increased by
movements of huge numbers of vehicles daily. The carbon dioxide increases the corrosion of reinforcing steel bars and affects residential structures by corroding plumb-
ing connections and support slabs. Regular maintenance through an integrated plan
is vital for economic reasons and will extend the life of a structure.

8.6 MAINTENANCE PLAN

The maintenance team may be responsible for many buildings. The team maintains
the reliability of the structures. This responsibility for bridges is clearly in the ministry of transportation.

For instance, in the United States the famous Golden Gate Bridge has a team that
continuously paints the bridge. After they finish they start again.

Concerning the time for performing the maintenance, it is complicated to calculate
the probability of failure as it is a matter for research rather than a practical issue.

The calculation of the structure probability of failure and the consequence of risk
is called the qualitative risk assessment. The other popular method is the quantitative risk assessment, and it can be handled through the maintenance team without
outside resources.

8.6.1 Assessment Process

The assessment of the concrete structure depends on the structure type, location, and
existing load and operation requirements. For risk assessment, the different critical
items, which affect economics, must be considered. The general definition of risk is
summarized in the following equation:

$$\text{Risk} = \text{probability of failure} \times \text{consequences}$$

So to define the risk assessment of a concrete structure, consider the factors that
may affect the economics of the business. The structural risk assessment is represented by the probability of failure and the consequence if the failure happened,

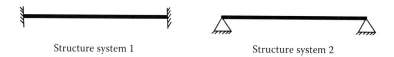

Structure system 1   Structure system 2

FIGURE 8.10   Structural redundancy.

noting that the failure may involve the whole building or part of it or only one concrete member.

The first step to define and calculate the quantitative risk assessment is to hold a
meeting with the maintenance team and maybe with a consultant engineering company in the case of huge structures.

This meeting will define the factors that affect the structure probability of failure,
such as corrosion on the steel reinforcement bars, new data that reveals concrete
strength lower than the design strength in the drawings and project specifications,
and the redundancy of the structure itself.

The required information about the corrosion of steel bars, concrete compressive
strength, and other factors can be obtained by collecting the data, performing the
visual inspection, and performing a detailed inspection using measurement equipment such as ultrasonic pulse velocity technique, rebound hammer, or other available techniques.

The factor that has the most influence on estimating the structure probability of
failure is the structure redundancy, which is explained in Figures 8.10 and 8.11.

If you need to design the beam and you receive two solutions, the first is a structure system to be fixed in the two ends and the second structure system is hinged at
the two ends, which system will you choose?

There are many factors that control your decision in selecting the suitable system.
Following are the advantages and disadvantages for these two systems:

Structural system 1:
  • The beam cross section will be smaller.
  • The connections will be big and complicated and have a shearing force
    and moment on them.
  • It is reasonable from architectural point of view.
  • The connection construction is complicated.
Structural system 2:
  • The beam cross section will be big.
  • The connection will be small as it is designed for shear force only.
  • It is easy to construct as the connection is simple.

Structural system 1   Structural system 2

FIGURE 8.11   Structural redundancy.

If you do not want to construct on site or maybe the construction will be carried out by the in-house engineers and workers, you will go toward the simple beam option. This type of thinking fails to see the maintenance point of view. Figure 8.11 illustrates the steps of collapse failure.

Structure system 2, which is the simple beam, assumes that you increase the load gradually. The beam can accommodate the load until a plastic hinge is formed in the middle of the beam at the point of maximum bending moment and then collapse will occur.

Conversely, for structure system 1, which is fixed at the two ends, when increasing the load gradually the weaker of the left or right connection will fail first. As shown in Figure 8.11, the plastic hinge will form on the left connection and increase the load and the other connection will be a plastic hinge, so it is now working as simple beam. Then by increasing the load, the third plastic hinge will be formulated and then collapse failure will occur.

From the above one can see that structural system 1 will take more time to collapse as it fails after three stages rather than structural system 2, which fails at the first stage, so the structure system 1 is more redundant than structure system 2.

Moreover, when you make a comparison between different reinforced concrete members such as slabs, beams, columns, and cantilevers, some members are more critical than others. As the cantilever is most critical, any defect will have a high deflection and then failure. When a column fails the load is distributed to other columns until failure of the whole building, so the column is low redundancy and very critical as any failure of it will fail the whole building; however, the cantilever failure will be a member failure only, and causes fewer consequences.

In the case of beam and slab, when you design a slab you assume a simple beam and calculate the maximum moment in the middle. Then you design the concrete slab by choosing the thickness and the steel reinforcement and the selected reinforcement will be distributed along the whole span, so theoretically by increasing the load the failure will be in point, but actually the surrounding area will carry part of this load so the redundancy of the slab is very high.

Some studies show that the reinforced concrete slab can accommodate load twice the design load. In some complicated structure systems it is important to use nonlinear methods in ultimate strength analysis, which in some textbooks is called pushover analysis, to determine the redundancy of the structure.

Conventional structural analysis practice relies on an idealized linear-elastic model to determine internal forces in components of the structure. Their adequacy is then determined by comparing the applied element forces with parametric code-check capacity formulas that are based on isolated component failure data.

In ultimate strength analysis, nonlinearities associated with the plasticity and large deformations of components under extreme load are included explicitly in the element modeling. The analysis tracks the interaction between components as member end restraints are modified and internal forces are redistributed in response to local stiffness changes. The sequence of nonlinear events leading to global collapse mechanisms and the associated system capacity are determined.

TABLE 8.6

Factors Affecting Probability of Failure

Structure	Redundancy 1–10	Age 1–10	Designer 1–10	Contractor 1–10	Code 1–10	Last Inspection 1–10	Total Score
S1	8	10	6	6	9	10	49
S2	5	2	7	3	3	10	30
S3	3	5	8	10	5	2	33
S4	7	4	2	5	5	1	24
S5	1	3	4	2	5	10	25
S6	3	1	4	5	6	7	26

Thus, while the typical linear design process checks for the adequacy of each individual component, the nonlinear ultimate strength analysis models the performance of the system as a whole.

Now, this nonlinear analysis capability exists for any structure analysis software package in the market. From this analysis one can determine how much more load the structure can carry than the design load until failure. In addition, we can know where the location of the first plastic hinge will formulate and from this one can know the critical member in the structure.

Therefore, for the concrete structure that requires a maintenance plan, the first step is to discuss all the structural parts that need to be inspected and repaired, which will be set in a table, and in this table the team will add all the factors that affect the structure's probability of failure.

Put a value on each different factor, for example, in the case of redundancy factor put a value from 1 to 10, for column it will be 10, and for slab it will be 2. the entire team must agree on these numbers based on their experience and repeat the same values or higher for other factors.

The other factors are the age, the engineering office that performs the design, and the contractor who constructs it, and also the code that is used in the design. As discussed before, in some countries seawater was used in mixing water. Obsolete codes allowed steel 6 mm diameter in stirrups, and they were completely corroded.

Table 8.6 contains the structure probability of failure factors, but when you use it make it match with your structure requirements. You must know that it is not limited to the factors only shown in the table.

Table 8.6 is a simple example of calculating the probability of failure.

Any meetings about probability of failure should include a discussion of calculated risk values. If you have had good experience with the design engineer, use lower values. If you have encountered engineering problems and question the engineer's competency, cite higher values. The same procedure applies to contractors. Consider the experience of your team and problems occurring with similar buildings around the country. For large or complicated structures, it may be worthwhile to hire a competent engineering firm to prepare a maintenance plan.

TABLE 8.7
Weights and Impacts

Structure	Impact on Person 1–10	Impact on Cost 1–10	Impact on Environment 1–10	Impact on Repetition 1–10	Total Score
S1	8	10	6	6	30
S2	5	2	7	3	17
S3	3	5	8	10	26
S4	7	4	2	5	18
S5	1	3	4	2	10
S6	4	5	3	5	17

TABLE 8.8
Risk Assessment Calculation

Structure	Probability of Failure	Consequences	Risk
S1	49	30	1470
S2	30	17	510
S3	33	26	858
S4	24	18	432
S5	25	10	250
S6	26	17	442

The next factor is inspection. If regular inspections are not performed, the building is a "black box" and you must assign the highest risk value. A structure that you know is in poor condition is less critical than a structure whose condition you know nothing about. Evaluate the consequences of failure by the same approach and include the economic impact of the failure of all or part of the structure.

Several hazard factors guide structure risk assessment including location, expenses, the impact of stopping production in an industrial facility, foundations supporting heavy machinery, and the results of a shutdown. Table 8.7 quantifies consequences on the same basis as calculating probability of failure.

Now, we have the quantity for the probability of failure and the consequences, so from Table 8.8 we can calculate the quantity risk assessment.

8.6.2   Risk-Based Inspection Maintenance Plan

After calculating risk assessment as shown in Table 8.8, classify the maintenance plan as a third of the risk structure colored red, the second third will be yellow, and the remaining third will be green. The propriety list will be as shown in Table 8.9.

So from Table 8.9, structures S1 and S3 are considered critical and where you will start inspection; structures S2 and S6 will be the second priority so they will

TABLE 8.9
Structure Priorities

Structure Priority	Color Code	Risk Values
S1	Red	1470
S3	Red	858
S2	Yellow	510
S6	Yellow	442
S4	Green	432
S5	Green	250

need to be inspected after the first two structures; and the last structures are S4 and S5, which present less risk so there is no need to inspect them at this time.

As shown above, this simplified method can be used to plan your maintenance for inspection and repair, taking into account that the budget is a very important factor in your plan. In the above example, the budget may be enough this year for inspecting structure S1 only, so you will have to plan for the following years using this table.

After you perform the inspection the data will be analyzed through the system as shown in the flowchart in Figure 8.12.

Nowadays, all the international organizations that have a number of different buildings and structure elements use the integrity management system, as it is very important to maintain every structure along its lifetime, taking into consideration all the historical data and the surrounding environmental conditions.

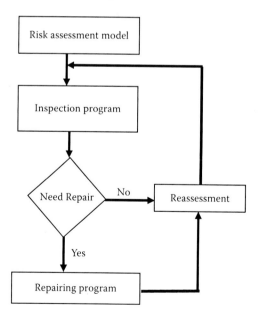

FIGURE 8.12 Risk assessment system flowchart.

The closed loop of structure risk assessment, inspection program, repair program, and reassessment is very important to maintain the structure in good condition and make a reliable match with the operations requirements.

The total quality control management system is very important in the case of reviewing design, construction, and maintenance to be matched with the operation requirements. In the case of proposing development or making an extension to a building or increasing the load, a system of management must be considered for changing the procedure to upgrade the previous risk-ranking table.

REFERENCES

Abd El-Kader, O., and A. Al-Kulaib. 1998. Kuwait Bridge management system. Eighth International Colloquium on Structural and Geotechnical Engineering, Cairo.

Arnon, B., S. Diamond, and N. S. Berke. 1997. *Steel corrosion in concrete.* E&FN Spon.

Berke, N. S., M. C. Dallaire, M. C. Hicks, and A. C. McDonald. 1996. Holistic approach to durability of steel reinforced concrete. In *Concrete in the service of mankind: Radical concrete technology,* R. K. Dhir and P. C. Hewelet, eds. United Kingdom: E&FN Spon.

Broomfield, J. P. 1997. *Corrosion of steel in concrete.* London: E&FN Spon.

El-Reedy, M. A., and M. A. Ahmed. 1998. Reliability-based tubular joint of offshore structure based inspection strategy. Offshore Mediterranean Conference, Italy.

Kamal S., O. Salama, and S. Elabiary. 1992. *Deteriorated concrete structure and methods of repair.* University Publishing House.

Khalil, A. B., M. M. Ahmed, and M. A. El-Reedy. 2000. Reliability analysis of reinforced concrete column. PhD thesis, Faculty of Engineering, Cairo University.

Morinaga, S. 1998. *Prediction of service lives of reinforced concrete buildings based on rate of corrosion of reinforcing steel.* Special report of the Institute of Technology, No. 23, June. Tokyo, Japan: Shimizu Corporation.

Raupach, M. 1996. Corrosion of steel in the area of cracks in concrete—laboratory tests and calculations using a transmission line model. In *Corrosion of reinforcement in concrete construction, special publication no. 183,* C. L. Page, P. B. Bamforth, and J. W. Figg, eds. Cambridge, UK: The Royal Society of Chemistry.

Tuutti, K. 1982. *Corrosion of steel in concrete.* Stockholm: Swedish Cement and Concrete Research Institute.

Index

A

ACI code
  ACI 207.4R (mixing speed), 203
  ACI 211.1, 164–168, 184
  ACI 211.2, 184
  ACI 214.77, 162*t*
  ACI 304.2R, 185
  ACI 318, 182
  ACI 5.11, 282
  capacity reduction factors, 89–90
  chloride content limits, 149, 150*t*
  Committee 305R (hot regions), 197
  concrete cover standards, 260
  concrete curing, 282–283
  concrete pouring standards, 270, 272, 274
  concrete strength standards, 163–164
  high-strength concrete definition, 205, 206
  hot regions guidelines, 197
  steel bar tolerances, allowable, 259*t*
  tolerances, allowable, 254*t*, 255
  wooden form code (ACI 318), 243
Admixtures
  accelerators, 192
  characteristics, 194*t*
  chemical, 166–168
  concrete strength, relationship between, 197*t*
  liquid, 192–193
  micro silica, 207
  mineral, 206–207, 209
  overview of usage, 191–192
  performance requirements, 196*t*
  powder, 192
  pumped concrete mix, used in, 185–186 (*see also* pumped concrete)
  retarder types, 192
  silica fume, 207, 208*f*, 209–210, 209*f*, 210
  testing (*see* Admixtures testing)
  water reducers, 192
Admixtures testing
  ash content analysis, 193–194
  chemical tests, 193
  chloride ion testing, 194
  control mixing, 195
  hydrogen number apparatus, 194
  performance testing, 195
  relative density, 194
AEA. *See* air-entraining agents (AEA), pumped concrete, use in

Aggregate testing
  acceptance and refusal limits, 119*t*
  bulk density, 126
  clay and fine materials testing, 123–124
  coarse aggregate specific gravity, 125–126
  fine aggregate test, 125
  Los Angeles test, 122
  mixing water test, 127
  on-site testing, 124
  percentage of aggregate absorption, 126–127
  sieve analysis test, 113, 115
  sieve shapes, 116*f*
  sieve sizes, 113–114*t*, 118*t*
  specific gravity test, 124
  volumetric weight, 126
  weight in sieves, 118*t*
Aggregates
  maximum size in concrete mixes, 165
  recycled (*see* recycled aggregate concrete)
  temperature control of, 202–203
  testing (*see* aggregate testing)
Air-entraining agents (AEA), pumped concrete, use in, 186
Al-Kulaib, 288
American Association of State Highways and Transportation Officials (AASHTO), 288
American Concrete Institute code. *See* ACI code
American National Standards Institute code (ANSI). *See* ANSI code
American Society for Testing and Materials (ATSM)
  C143-90a, 177
  C192-76, 172
  cement limits, 104
  cylinder test standards, 172
  grading requirements for coarse aggregate, 120
  high-strength concrete codes, 206
  slump test, 177, 179
ANSI code
  live load, 63, 64, 67
  load parameters, 58–59
  typical live load statistics, 59*t*
  wind load, 68
Arithmetic mean, 154
ASTM. *See* American Society for Testing and Materials (ATSM)